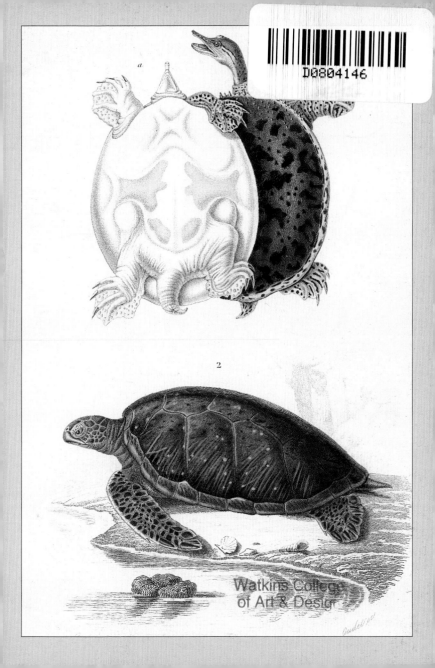

This dauw, a mountain zebra, is less evenly striped than most of its close African relatives. Darwin believed that the stripes that often appear on the shoulders, back and legs of the Equidae and several other sub-groups of mammals were an indication that they were descended from a common ancestor.

*One of the most remarkable products of artificial selection
by English stockbreeders in Darwin's time: the shorthorn cow.
It was the result of a long process of improvement that
combined the selection of individual variations
with closely monitored breeding.*

Below, the female of the death's-head hawkmoth (Acherontia atropos L.), *with wings outspread. To human eyes, the design on the upper part of its back seems to resemble a skull, as if designed to scare away a potential enemy.*

Opposite, a collection of beetles. The central pair of stag beetles display a marked sexual dimorphism. The powerful mandibles of the male (below) are used in combat.

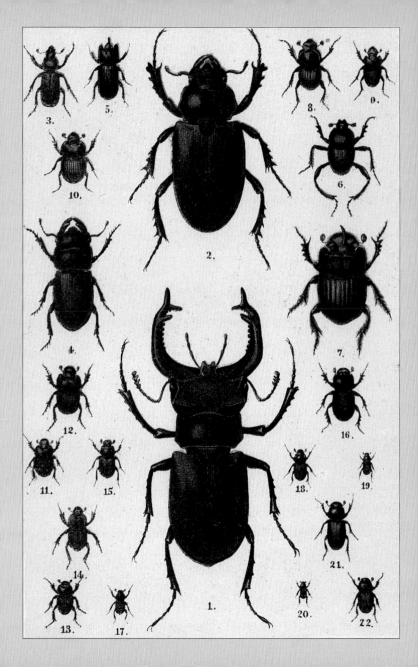

*Above, the Belgian carrier pigeon. Opposite,
this lesser bird of paradise (Paradisea minor),
painted by John Gould, is a spectacular example of
the external differences between the sexes that develop
through sexual selection. The male's display plumage
plays a central role in attracting a mate, but has
grown so large that it has almost become a handicap.*

Above, one of the several thousand species of orchids,
which were one of Darwin's true scientific passions.
Climbing plants were another great interest;
opposite, a European dodder, its tendrils covered
with suckers, entwines itself around garden
plants and feeds from them.

CONTENTS

DARWIN
AND THE SCIENCE
OF EVOLUTION

Patrick Tort

DISCOVERIES®
HARRY N. ABRAMS, INC., PUBLISHERS

'When I left the school I was for my age neither high nor low in it; and I believe that I was considered by all my masters and by my father as a very ordinary boy, rather below the common standard in intellect. To my deep mortification my father once said to me, "You care for nothing but shooting, dogs, and rat-catching, and you will be a disgrace to yourself and all your family".'

Autobiography, 1876

CHAPTER 1

AN UNCERTAIN BEGINNING

Charles Darwin (1809–82), aged thirty-one, in a watercolour by George Richmond, 1840. Right, his single-lens microscope, an instrument which he said no naturalist should be without.

Family expectations

The naturalist who, from 1859 onward, was to become the world's most famous evolutionary theorist owed this in part to a very distinctive sense of family. When he was born on 12 February 1809, at Shrewsbury in Shropshire, Charles Darwin was already surrounded by examples to follow. On the side of his father, Robert Waring Darwin – a doctor in the town since 1787 and a remarkably young fellow of the Royal Society (1788) and the Medical Society – he had the field of medicine set before him as a model. Medicine had also been the profession of his grandfather Erasmus Darwin, possibly one of the most original figures in eighteenth-century Europe. His book *Zoonomia* of 1796 pre-dated by four years the work of French scholar Jean-Baptiste Lamarck in clearly rejecting the doctrine of creationism and the fixed categorization of living things. This eminent scholar was one of the first to imagine that organisms and species could be gradually transformed under the influence of their needs.

On the side of his mother, Susannah, who died from illness while Charles was a boy, he was descended from the Wedgwoods, a line of industrial entrepreneurs. The family patriarch was Josiah Wedgwood, the rebellious, partially handicapped and relentlessly innovative son of a family of potters, who revolutionized the art of decorative ceramics and was a pioneer of industrial aesthetics. With a tradition of non-conformism on both sides of the family, Charles too found himself taking a radically different path so that his own talents could be fulfilled.

Erasmus Darwin (1731–1802; left), was Charles's paternal grandfather. A doctor, physicist, geologist, naturalist, inventor and poet, he was progressive both in politics (he supported the American and French revolutions, and wanted to extend the right to vote and to abolish slavery) and in philosophy. An opponent of finalism, he was later accused of atheism, and his *Zoonomia* was placed on the Index of Forbidden Books. His best friend was Josiah Wedgwood (1730–95; opposite), prince of ceramicists, ceramicist to princes and Charles's maternal grandfather. He founded the model industrial village of Etruria in 1769, and married his cousin Sarah in 1764.

Connections and influences

There were very strong bonds between the Darwin and Wedgwood families, and the first of these was a deep friendship between the two patriarchs. Erasmus the scholar and Josiah the entrepreneur were united in their opposition to the dominant ideas in contemporary politics and religion. They had little faith in God, took a passionate interest in scientific discoveries, loved democracy, were opposed to slavery, and possessed the energy needed to put their convictions into practice.

The second bond, founded on the dearest wish of these two friends, was the alliance in marriage of their children. Robert Darwin married Josiah's daughter Susannah in 1796, and Charles was once again following a precedent set by his father when he married Josiah's grand-daughter Emma in 1839. The marriage took place shortly after Charles's return from almost five years away from his family on a voyage that provided confirmation both of his theories on the

Above, a portrait of the Wedgwood family by George Stubbs, *c.* 1780. From right to left: Josiah and his wife Sarah (seated), John, Josiah II, Susannah, the future mother of Charles, Catherine (standing), Thomas, Sarah Elizabeth and Mary Ann.

gradual transformation of living things within their environment, and of his own decision to become a naturalist.

It is already apparent how Charles may have come to understand that the secret underlying the development of the living world lay in finding a compromise between resemblance to ancestors and individual variation, evolutionary divergence, the promotion of innovation. In June 1817 his mother died, and Charles grieved for her deeply.

Early schooldays

Charles spent a year as a day-boy at the school of the Reverend G. Case, minister of the Unitarian Chapel on Shrewsbury High Street. Unitarianism was the religion of the Darwin family, and was in itself the embodiment of dissent with the Church of England. Its central tenet was the rejection of the Catholic doctrine of the Trinity (the division of God into

the Father, Son and Holy Spirit), and this was accompanied by a universal attitude of tolerance, openness and charity. For Charles this was the start of a half-hearted and, by his own admission, rather inglorious schooling, but one which nevertheless led to an increasing passion for studying nature and collecting. The following year, 1818, he became a boarder at the 'big school' of Dr Sam Butler, in the same town. He spent seven years there, and they left somewhat dreary memories, obviously linked to the difficulties he had in learning the strictly classical subjects that made up the curriculum: ancient languages and literature, history and geography. His only interests were hunting and fishing (which he gave up in his mature years for humanitarian reasons), as well as an insatiable appetite for observing plants, insects and birds. Slowly, however, he began to read poetry, learned to appreciate the beauty of landscapes, and practised geometry and experimental chemistry. In some of these subjects, particularly the latter, he followed the tastes of his brother Erasmus Alvey, who was five years older.

A path of his own

Erasmus was Charles's childhood ally and secret protector, and throughout his life conveyed to the world the image of a dilettante with frivolous tastes, unproductive inclinations and knowledge that he made little use of; he had qualified as a doctor, but never practised. However, he was a secondary figure in Charles's childhood in comparison with the much more central figure of their father, whose corpulence and good-hearted but serious nature

• The school as a means of education to me was simply a blank. •
Autobiography
[Opposite, the library at Shrewsbury School, engraving, 1843.]

Charles and Emily Catherine (Catty), fifteen months his junior, were the two youngest children in the Darwin family (left, in 1816, in a pastel drawing by Rolinda Sharples). Catherine's affection for the closest of her brothers can be seen in the letters she wrote to him during the voyage of the *Beagle*, as well as in her wish for him to become a parson and marry Frances (Fanny) Wedgwood, who was to die of illness in 1832. However, Charles did not join the church, and eventually married the youngest of the Wedgwood sisters, Emma. It seems that the overbearing personalities of Charles's elder sisters rubbed off on the youngest. Despite her fragile health, she had the temperament of a cantankerous governess, which was disliked but patiently tolerated by Charles and his children.

inspired not only respect and admiration, but also fear and an anxiety not to disappoint.

Charles loved his father, but from a distance that could only have been reduced had he fulfilled Robert Darwin's hope that his son would take up a medical career. Writing later in life, he recalled his father's excellent memory, his generosity, his capacity for intuition, and the 'power of sympathy' which earned him the affection and trust of his patients. He had less to say about his demanding nature, but it was this trait which led to Charles's growing anxiety as his studies progressed, since he was acutely aware of the family expectations which he could not satisfy, and which his elder brother had, skilfully and cynically, first fulfilled and then dashed. By distinguishing himself as a doctor, Charles would have made solid his family connection to an absent figure whose promise could have been reborn and realized in the next generation: the first Charles Darwin (1758–78), eldest son of Erasmus and admired brother of Robert, had died at the age of 20 in Edinburgh, of septicaemia caused by an accidental wound which occurred while performing an autopsy.

Anyone seeking a psychoanalytical explanation of the determining factors behind Darwin's career and thought might note that the young Charles Darwin was named after a dead man. His namesake, that late and exceptional uncle, had carried the highest hopes of the great Erasmus, who held up the late Charles as an example to his third son, Robert, who, although at first hating medicine and operations, nevertheless fulfilled his own father's wish. Doubtless he did this vicariously and submissively, and with a deep sense of constant effort; this may have made him even stricter when he came to pass on symbolically to his own son Charles, not only the Christian name of the deceased

Charles's father, Robert Waring Darwin (1766–1848; below), was the fourth child and third son of Erasmus Darwin. In 1785, at the age of nineteen, he published a study of the persistence of coloured patches on the retina after viewing bright objects. A good but demanding man, he had some harsh words for Charles's carelessness as a schoolboy and a student. Later, in his *Autobiography*, Charles admitted to having relied rather heavily on his father's money to avoid having to practise medicine. However, the University of Edinburgh (engraving, opposite) turned him away from this path for ever.

but also the task of taking the place of the brilliant young man whom family legend had given something of an aura of martyrdom. The situation was severe enough for Darwin to begin to feel absent from his own childhood, and robbed of his own choices by a story that was not his own. His reverence for his grandfather was closely combined with a wish to

•Dr Munro made his lectures on human anatomy as dull as he was himself, and the subject disgusted me. It has proved one of the greatest evils in my life that I was not urged to practise dissection, for I should soon have got

forget the whole business, and so he refused to take up the field of medicine which had not been able to cure his mother. It would take time before he found the path of compromise between the paradoxes of filial obedience, and the indecisiveness that characterized his time at university was almost certainly due to the pressure of his family's wishes. As a result, Charles was burdened with expectations that he would have to disappoint in order to begin to live for himself. It is no surprise, then, that Charles's medical studies were disappointing.

over my disgust; and the practice would have been invaluable for all my future work. This has been an irremediable evil, as well as my incapacity to draw.•
Autobiography

Edinburgh: a brush with Lamarck

The two years spent at the University of Edinburgh, where Charles enrolled on 22 October 1825, were filled with a large measure of boredom and no small amount of repulsion at the dissections which had caused the death of his namesake uncle. He wrote that his only source of interest was the lectures of Thomas Charles Hope, the chemist who, twenty years earlier, had determined that water reaches its maximum density at a temperature of 4 degrees Celsius. Following pursuits of his own, Charles befriended some young naturalists whose leader seems to have been Robert Edmond Grant (1793–1874), some sixteen years his senior. Grant was a passionate champion of the theories of Lamarck: species are affected by changes in the environment that alter the needs of organisms. These then adapt in response, and those adaptations can be inherited by their descendants. Despite the proselytizing of his young teacher, Darwin displayed no more conscious attention to these ideas at this time than he showed for those of his grandfather Erasmus, although he had read and admired *Zoonomia*.

In 1826, his undivided interest in the natural sciences led him, at Grant's instigation, to deliver two short papers to the Plinian Society, one on the

ciliary movement of Flustra larvae, and the other on the eggs of the wormlike *Pontobdella muricata*. He attended several academic societies, and was taught the art of stuffing birds by a black taxidermist. During the summer or autumn, he would visit Maer Hall, home of his uncle Josiah Wedgwood II, the son and successor of the great ceramicist. There Charles could enjoy nature, hunting and a more carefree environment than the austere home of his father. Robert Darwin had been obliged to give up the dream of seeing his son take up a medical career, and instead had agreed with him upon an alternative career which doubtless seemed less brilliant, although it was respectable and relatively secure: the modest life of a country parson. This prospect would nevertheless provide the young Darwin with enough free time for his research as a naturalist.

Cambridge: theology, beetles and botany

It was with a minimum of resolve that Charles enrolled at Cambridge University on 15 October 1827. At home with a tutor, he caught up with the classical studies that he had forgotten, before joining the university early the following year for a three-year stay.

At Cambridge, he had little interest in most subjects, although as he prepared for his degree, he did enjoy studying Euclidian geometry and natural theology. The latter discipline, which has a long

•But no pursuit at Cambridge was followed with nearly so much eagerness or gave me so much pleasure as collecting beetles. It was the mere passion for collecting, for I did not dissect them and rarely compared their external characters with published descriptions, but got them named anyhow.•

Autobiography

This lasting passion is evident in Darwin's meticulous beetle classifications (opposite), the witty sketch by a fellow student, Albert Way, showing Charles on a giant beetle (below), and this specimen brought back from the Falkland Islands (left).

DARWIN & his HOBBY.

history in Europe, was represented in England by the works of William Paley. It consisted of an admiring description of the harmony of nature, which could ultimately be explained only through divine providence bringing order to the universe, following laws that were intended both to express its own perfection and to have it recognized by humankind, its principal beneficiaries within creation.

The epitome of an apologist discourse, its aim was to interpret scientific knowledge and reshape it as a series of 'proofs' that would reinforce belief in the providence of an omniscient creator. Natural theology was the guarantee of ecclesiastical influence keeping its grip on thought, and the objective bastion of dogmatic theology, although it did play a part in moderating the latter in order to avoid any major confrontations with science. In striving to keep the Church in control of the sciences (whose heavy emphasis on mechanical structure tended increasingly to distance

At Christ's College, Cambridge (below), Charles did as little work as possible. He showed far more interest in art than in his studies, and led the rather dissolute life of a well-off student, socializing and having fun.

it from any concept of transcendent plan and design), natural theology would go through the same procedure with any new object presented for its evaluation. Faced with any series of interconnected facts or phenomena in 'harmonious' combination, it would put forward arguments that denied the possibility of a causal explanation based only the operation of 'blind' nature. Consequently, any phenomenon was made a homage to the providential wisdom and will of God, which alone were capable of creating the happy and beautiful appropriateness of its inner structure and its assigned place in the universe. The balance of parts and the harmony of reponses to the external world apparent in every living creature could not be explained by anything so incredible as 'chance', but only as a plan prearranged by divine will.

This type of reasoning is called teleology (explanation by ends or 'final causes'), and its intention is to promote the existence, omnipotence and superior intelligence of God by arousing admiration for the 'wonders of Nature'. Teleology, also called finalism, was dealt a mortal blow by Darwin scarcely ten years later, when he reached his own understanding of the strictly immanent reasons for the equilibria of the living world. However, his studies of Paley did enable him to explore and savour what genuinely interested him: nature's infinite resources for combination, the fine adaptations of living things, the interdependence of organs and the harmony in which they work together.

William Paley (1743–1805; below), Anglican archdeacon and author of *Principles of Moral and Political Philosophy* (1785), *Evidences of Christianity* (1791) and *Natural Theology* (1802), was at that time considered an authority: 'I did not at that time trouble myself about Paley's premises; and taking these on trust I was charmed and convinced by the long line of argumentation.' (*Autobiography*). Paley had also been a fierce critic of the transformist theories of Erasmus Darwin's *Zoonomia*.

At around the same time, Darwin collected and meticulously classified a great number of beetles, his work attracting the attention and respect of several specialists. He became familiar with botany through the lectures of John Stevens Henslow (1796–1861), whom he accompanied on walks; he also frequented Henslow's soirées, which were enlivened by the presence of brilliant figures from the intellectual world, including William Whewell, a philosopher, theologian, mathematician and theoretician of science; and the naturalist Leonard Jenyns. He studied Adam Smith and John Locke, the founder of English empirical philosophy, and read with fascination the tale of the *Voyage* of Alexander von Humboldt (1769–1859), a story that stayed with him throughout his life. He also read and admired *A Preliminary Discourse on the Study of Natural Philosophy* by the astronomer John Herschel.

Discovering geology

In January 1831, Darwin was rewarded for his labours, obtaining a Bachelor of Arts degree without honours, with the very respectable placing of tenth out of the 178 candidates admitted. In response to friendly encouragement from Henslow, the young Charles, now aged twenty-two, decided to study geology, a subject which had caught his interest and which had been recently placed in a new light by the publication in 1830 of the first volume of Charles Lyell's *Principles of Geology*. At Cambridge, Darwin had not attended the classes of Adam Sedgwick, then president of the Geological Society and Woodwardian Professor, although had he done so, they might have made him pay more attention to this discipline, as well as to the importance of fossil evidence as a clue to the relative simultaneity of deposits found in different regions.

This omission was soon to be rectified through Henslow's intervention. The latter used his influence to make his colleague Sedgwick accept Charles, then

M aer Hall (above), in Staffordshire, was the home of Josiah Wedgwood II, his wife Bessy and their family. 'Uncle Jos' had succeeded the elder Josiah as head of the pottery works, and his support was a great help to Charles in obtaining the good will and understanding of his father, who also helped the Wedgwoods with the management of their finances. Charles loved Maer and every year returned with the greatest pleasure to this forested paradise for the hunting season.

Darwin missed the geology lectures of Adam Sedgwick (below) at Cambridge, but went on to befriend him in 1831. Soon afterwards, impressed by the scientific work Darwin did during his travels, Sedgwick predicted he would have a brilliant future as a naturalist, although in 1860 he became an opponent of Darwin's 'transformist' theory. Charles later expressed both sympathy and some annoyance with this geologist and clergyman who had been unable to cast off the scientific dogma that weighed heavily on his discipline.

dreaming of a voyage to Tenerife, as a companion on a geological field trip to north Wales. Sedgwick spent an August night at the Darwins' house in Shrewsbury, and made a great impression on Charles. The trip, which began the next day, was intended to teach the young man 'how to make out the geology of a country'. However, as Darwin admitted forty-five years later in his *Autobiography* (1876), neither of them noticed the clear evidence of glacial action that had marked the valley of Cwm Idwal. Charles later left Sedgwick, travelled to Barmouth and then Shrewsbury, and managed to reach Maer in time for the start of the partridge hunting season.

'If a person should ask my advice, before undertaking a long voyage, my answer would depend upon his possessing a decided taste for some branch of knowledge, which could by such means be improved.... It is necessary to look forward to a harvest, however distant it may be, when some fruit will be reaped, some good effected.'

Autobiography, 1839

CHAPTER 2

THE VOYAGE OF THE *BEAGLE*

The *Beagle* (painted by John Chancellor on 17 October 1835 at James Island in the Galapagos) was a three-masted brig with ten cannons. Rather unstable in rough weather, it was about 30 m (100 ft) long by 8 m (26 ft) wide. There were 74 people aboard, and very little space. Darwin recorded the events and observations of the voyage in notebooks (right) which became the raw material for his *Journal of Researches*.

The *Beagle* and its captain

At Shrewsbury, Charles found a letter from Henslow informing him that Captain FitzRoy was offering to share a cabin on his sailing ship, the *Beagle*, with a young naturalist who would be an unpaid volunteer on a voyage to circumnavigate the southern hemisphere, with the intention of completing a hydrographic survey of the coasts of South America and carrying out chronometric readings. Charles wanted to accept, but met with the objections of his father, who added: 'If you can find any man of common sense who advises you to go, I will give my consent.' Charles was resigned, and wrote a letter of refusal.

1. *Mr. Darwin's Sea*
3. *Mr. Darwin's Cha*

The following day, he went to Maer where he told the whole tale, and his 'Uncle Jos', a man of good judgment, came back to Shrewsbury the next day to plead Charles's cause. It was then that Robert Darwin gave his wholehearted consent.

Charles went to Cambridge, where he saw Henslow, and then to London, where he met the young, elegant, religious, aristocratic and authoritarian Captain FitzRoy, then 26 years of age, whose turbulent and unstable temperament, proud and domineering nature, quick temper and conservative beliefs often made their relationship a difficult one. Darwin wrote in 1876 that FitzRoy, a follower of the Swiss physiognomist Lavater, later admitted to having almost rejected Darwin because of the shape of his nose.

His presence having been squared with the Admiralty, Darwin left England on board the *Beagle*

Captain Robert FitzRoy (above), was the son of a general and descended from the Stuarts. A Tory and a strict Christian, he was also an advocate of slavery.

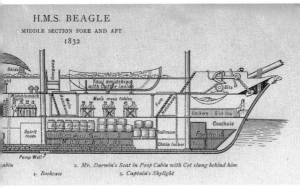

H.M.S. BEAGLE
MIDDLE SECTION FORE AND AFT
1832

2. Mr. Darwin's Seat in Poop Cabin with Cot slung behind him
4. Bookcase   5. Captain's Skylight

FitzRoy was a good captain and an affable host for the 1741 days of the voyage. However, in March 1832 he lost patience at the cheek of Darwin, a Whig abolitionist, when the latter doubted that a black slave from Brazil, summoned by his master, could sincerely testify to his own happiness. By 1839 he had become rather jealous of Darwin's scientific success. Appointed Vice-Admiral in 1863 after a political and administrative career, he committed suicide in 1865, still hostile to Darwinian theory and deploring the collapse of the Confederate states of the southern USA. Left, a cross-section of the *Beagle*, 1832. Below, Darwin's sextant.

on 27 December 1831, with his naturalist's equipment and a carefully chosen scientific library, which notably included the first volume of Lyell's *Principles of Geology*. He came to consider the voyage as 'the most important event' of his life.

Volcanic islands

On 6 January 1832 the *Beagle* arrived at Tenerife, in the Canary Islands. Darwin had to give up his dream of knowing and exploring these islands, since the authorities, fearing cholera, would not allow any disembarkation. He had to wait a further ten days before he could visit a volcanic island, when the ship arrived at Praia on São Tiago, Cape Verde. There Darwin attested the truth of Lyell's theory (partly inherited from Hutton, the Scottish founder of the new geology) that these islands were formed by an uplift followed by a gradual collapse or subsidence around the craters, with this balancing dynamic in the movements of the earth seeming to follow a general rule of compensation.

These observations were confirmed in February by examination of the rocks of São Paulo, and later on the island of Fernando de Noronha. On each landfall, Darwin made many physical, climatological, zoological,

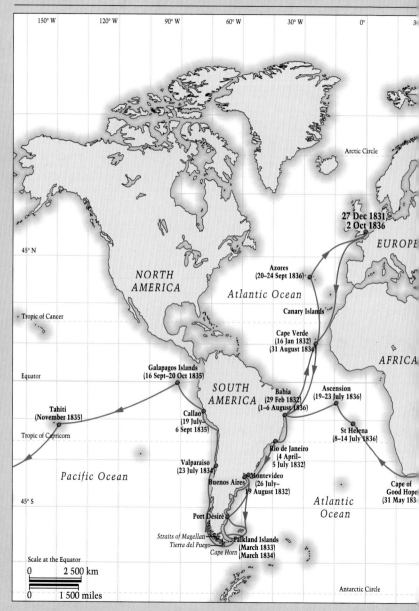

150° W   120° W   90° W   60° W   30° W   0°   3

Arctic Circle

27 Dec 1831
2 Oct 1836

EUROPE

Azores
(20–24 Sept 1836)

45° N

NORTH
AMERICA

Atlantic Ocean

Canary Islands

Tropic of Cancer

Cape Verde
(16 Jan 1832)
(31 August 1836)

AFRICA

Galapagos Islands
(16 Sept–20 Oct 1835)

Equator

SOUTH
AMERICA

Bahia
(29 Feb 1832)
(1–6 August 1836)

Ascension
(19–23 July 1836)

Tahiti
(November 1835)

Callao
(19 July–
6 Sept 1835)

St Helena
(8–14 July 1836)

Tropic of Capricorn

Rio de Janeiro
(4 April–
5 July 1832)

Valparaiso
(23 July 1834)

Pacific Ocean

Montevideo
(26 July–
19 August 1832)

Buenos Aires

45° S

Atlantic
Ocean

Cape of
Good Hope
(31 May 183

Port Désiré

Scale at the Equator

Straits of Magellan
Tierra del Fuego

Falkland Islands
(March 1833)
(March 1834)

Cape Horn

0   2 500 km

0   1 500 miles

Antarctic Circle

‘The voyage of the *Beagle* has been by far the most important event in my life and has determined my whole career.... I have always felt that I owe to the voyage the first real training or education of my mind. I was led to attend closely to several branches of natural history, and thus my powers of observation were improved, though they were already fairly developed. The investigation of the geology of all places visited was far more important, as reasoning here comes into play. On first examining a new district nothing can appear more hopeless than the chaos of rocks; but by recording the stratification and nature of the rocks and fossils at many points, always reasoning and predicting what will be found elsewhere, light soon begins to dawn on the district, and the structure of the whole becomes more or less intelligible.’

Autobiography
[Left, map of the world showing the *Beagle*'s principal ports of call on its five-year voyage.]

botanical and anthropological observations, collecting, observing and noting everything, from the the dust in the atmosphere at sea – samples of which he sent to be studied by Ehrenberg, the German specialist in microscopic lifeforms – to cultivated plants, domestic animals, and details of the daily life of the human populations.

Brazil: splendour and slavery

Having reached first Bahia, then Rio in Brazil, he undertook, from 8 to 23 April 1832, an overland trip during which he marvelled at the tropical landscape, observed the birds, parasitic plants, orchids, coffee growing, the use of manioc as food, animal herds, game, cooking, climbing plants, tree ferns, and insects. He carried out an experiment on the total regeneration of a worm, and studied the attraction of a beetle for a mushroom, a phenomenon he had already seen in England. He also witnessed the distressing spectacle of slavery, and felt a horror for it which he carried with him for the rest of his life. Indeed, this supposed 'domestication of man' was always a source of repulsion in Darwin's eyes, and his memories of slavery were later at the source of his reservations regarding the recognition of Man as the 'most domesticated species'.

Uruguay, Argentina: excavations, observations and another brush with Lamarck

The *Beagle* continued its voyage and arrived in Uruguay at Montevideo, which became the departure point for several expeditions into nearby Argentina.

'I was very nearly being an eyewitness to one of those atrocious acts, which can only take place in a slave country. Owing to a quarrel and a lawsuit, the owner was on the point of taking all the women and children from the men, and selling them separately at the public auction at Rio. Interest, and not any feeling of compassion, prevented this act.... I may mention one very trifling anecdote, which at the time struck me more forcibly than any story of cruelty. I was crossing a ferry with a Negro, who was uncommonly stupid. In endeavouring to make him understand, I talked loud, and made signs, in doing which I passed my hand near his face. He, I suppose, thought I was in a passion, and was going to strike him; for instantly, with a frightened look and half-shut eyes, he dropped his hands. I shall never forget my feelings of surprise, disgust, and shame, at seeing a great and powerful man afraid even to ward off a blow, directed, as he thought, at his face.'

Journal of Researches,
14 April 1832

There, at Punta Alta, Darwin discovered a jawbone containing a tooth from *Megatherium*, a gigantic fossil edentate mammal (the Edentata, like the present-day sloth [*Bradypus*] do possess teeth, but they are small and few in number). In the region of Maldonado, Uruguay, Darwin collected many birds and reptiles. He shared the tough life of the gauchos, observed flocks of rheas, studied the vegetation, climate and agriculture, and rodents including the capybara or water cavy, and he speculated about the blindness of the tuco-tuco (*Ctenomys brasiliensis*), which closely resembles a mole. He took a particular interest in the geographical distribution of birds, as well as the physical and behavioural similarities between otherwise distant species.

It was between 14 and 27 November, during the *Beagle*'s stay at Montevideo, that Darwin received the second volume of Lyell's *Principles of Geology*, which discussed and refuted Lamarck's 'transformist' theory.

Above, the toucan. Its huge bill is light because it is made up of cavities separated by a tissue of bony fibres. Opposite, *Trogon viridis*. Below, an engraving that depicts a master beating a slave.

An unlucky predecessor: Jean-Baptiste Lamarck

It took some time for Darwin to recognize the talent of this great French naturalist, who coined the term 'biology' in 1802, was a pioneer in meteorology and invertebrate palaeontology (as well as of the distinction between vertebrate and invertebrate animals), and was the most important of his predecessors in the field of 'transformism'. Originally a botanist, Lamarck became a member of the Académie des Sciences in 1779, and in 1793 joined the Muséum National d'Histoire Naturelle, taking charge of 'Insects and Worms', and becoming a specialist in invertebrates. Through his work of classifying invertebrates, he declared himself a 'transformist' in 1800. Time and circumstances influence organisms, stimulating needs and modelling habits that develop and strengthen their organs and structures, which then change and diversify. These modifications are passed on to subsequent generations. It is the instigatory action of the environment on the needs and habits which adapts and modifies the form of organs, and hence of organisms and groups, making probable the long and branching passage from the simplest microscopic lifeforms, which he believed were spontaneously generated under certain chemical conditions, to the most complex, namely man and the vertebrate mammals (*Discours d'ouverture du Cours de l'An VIII*). A year later, in 1801, in *Système des animaux sans vertèbres*, he recognized the resemblance between fossil and present-day species, and criticized the theory of catastrophes that was extolled,

Jean-Baptiste Pierre-Antoine Monet de Lamarck (1744–1829; below) was the first naturalist to make 'transformism' the central issue of his work. A botanist, zoologist, meteorologist and something of a philosopher, he was fiercely opposed by Cuvier who, on his death, wrote a eulogy so vindictive that the Academy had to turn it down. Lamarck went blind in 1820, but managed to complete the seven volumes of his *Histoire naturelle des animaux sans vertèbres* with the help of his two daughters. He died poor and was buried in a paupers' grave at Montparnasse.

though he avoided naming him, by Georges Cuvier, a powerful colleague who championed creationism and the concept of 'global revolutions'. In *Discours d'ouverture du Cours de l'An X* (1802), Lamarck emphasized the need to go beyond classification and understand the processes by which species are formed, transformed and 'perfected'.

Below, a table from Lamarck's 1815 *Histoire naturelle des animaux sans vertèbres*, showing the order in which he believed animals had come into being. It was reworked and completed in 1820 in the *Système analytique*

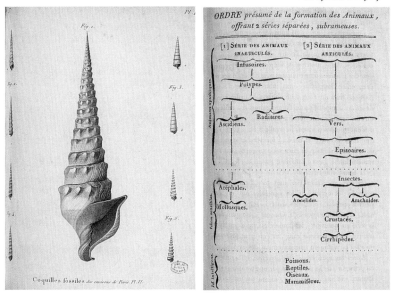

Coquilles fossiles *des environs de Paris. Pl. II.*

ORDRE présumé de la formation des Animaux, offrant 2 séries séparées, subrameuses.

Like Darwin, he used the modifications made to animals and plants through human domestication as a basis for inferring nature's own power to transform. In 1809, he published *Philosophie zoologique*, in which he synthesized the principles of 'transformism'. A victim of poverty and of the inflexible power of Cuvier, against which the remarkable Etienne Geoffroy Saint-Hilaire was later to make a stand, Lamarck died poor and blind, leaving his successors to inherit a doctrine that is often condensed to two main ideas: the principle of use and disuse (the frequent use of an organ strengthens and improves it;

des connaissances positives de l'homme, where it was presented as a tree diagram, a format later adopted by many 'transformists'. Note the close relationship between the Cirripedia and the Crustacea, which was definitively confirmed by Darwin about thirty years later. Left, a plate showing fossil shells.

a decline in use weakens it and tends to make it disappear), and the principle of the inheritance of acquired characteristics (each alteration, whether an acquisition or a loss, that occurs under the influence of the environment during the life of an individual is passed on to its descendants).

Lyell and uniformitarianism

Lyell was still a long way from his later conversion to 'transformism', and he only presented Lamarck's work in order to refute it. But the refutation was done respectfully, approving the iconoclastic spirit which Lyell more than shared. This text was the first to use the word 'evolution' in its modern sense, and went on to lead many of Lyell's English readers towards the idea of the transmutation of species. It is pleasing to recall that it was Henslow who, although himself convinced that the history of the Earth was marked by a series of cataclysmic events, advised Darwin to take the first volume of Lyell's work on his voyage, and in fact it was personally given to him by the austere FitzRoy, the very personification of Biblical orthodoxy. Henslow had tempered his advice with a recommendation: to study the book without believing a word of the doctrine it championed. It is also interesting to note that Lyell made a similar gesture within the book itself, by publicizing Lamarck's work even while doing his utmost to oppose him. It could even be suggested that this

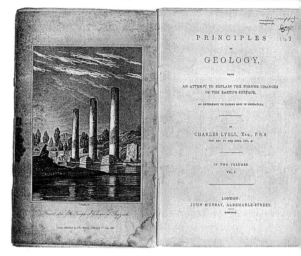

Frontispiece and title page of the first volume of the first edition of *Principles of Geology* by Charles Lyell (1830; opposite). The second volume, in 1832, contained the famous critical analysis of Lamarck, and Darwin received a copy during a stopover at Montevideo.

•The science of geology is enormously indebted to Lyell – more, as I believe, than to any other man who ever lived. [...] I am proud to remember that the first place, namely St Jago, in the Cape Verde Archipelago, which I geologised, convinced me of the infinite superiority of Lyell's views over those advocated in any other work known to me.•

Autobiography

continuous, ambiguous, multiple mediation had a positive influence on Darwin.

It was within the field of geology that Darwin first espoused the new ideas. In England, Charles Lyell was the most famous representative of the Uniformitarian geology, an adaptation of actualism or the theory of 'present causes' proposed by several European geologists. Changes in the earth's geology are not, as was believed by Cuvier and the Christian 'catastrophists', due to cataclysmic events. The natural events which can today be observed to affect geological processes are the same ones which have always modelled the Earth's surface. The first promoters of this theory were the German Karl von Hoff (1822), and the Frenchman Constant Prévost who directly influenced Lyell and whose anti-catastrophist theories date from the same period. When Darwin later observed the transformation of domestic animals in the present day, and deduced the idea of an analogous process occurring in Nature since the appearance of the earliest living forms, he was taking a similar 'actualistic' approach.

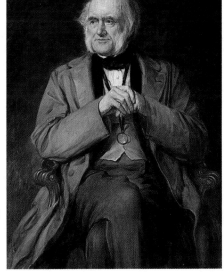

Uniformitarianism was Lyell's theory that major geological effects over a long timespan are produced through the slow accumulation of small-scale effects that can be seen within a human lifetime. Lyell did not realize that this theory established the conditions necessary for natural phenomena to evolve. Sedgwick feared this and in 1831 he warned Lyell of the risk of a drift in his meaning. Later, Lyell's support of Darwin and belief in his theories confirmed this implied coherence in the two doctrines.

Tierra del Fuego: the 'savage' and the 'educated'

The next stopover, at the southern extremity of Argentina, was Tierra del Fuego. The *Beagle* sailed along its coasts between 16 December 1832 and 28 February 1833. The party sent ashore by FitzRoy on 18 December was met with astonishment but without hostility by a group of native Fuegians who stood on the beach.

It was then that Darwin assessed 'how wide was the difference, between savage and civilized man. It is greater', he wrote in his *Journal of Researches*, 'than between a wild and domesticated animal, in as much as in man there is a greater power of improvement.' He was speaking as a young middle-class Englishman, with the unthinking ethnocentrism of an educated member of the nation which then believed itself to be the most civilized in the world, on seeing for the first time a race who had had no prior contact with Europeans and who were living a traditional existence on this wretched part of the American continent, in the mist and the cold, and who, apart from their native language, knew only a few mixed-up scraps of Spanish and Portuguese. Almost forty years later, on the last page of his great book *The Descent of Man* (1871), Darwin described this shock again: 'The astonishment which I felt on first seeing a party of Fuegians on a wild and broken shore will never be forgotten by me, for the reflection at once rushed into my mind – such were our ancestors. These men were absolutely naked and bedaubed with paint, their long

On board the *Beagle* were three Fuegians – Jemmy Button, Fuegia Basket and York Minster (above). After three years in England, they were being taken back to their homeland, indoctrinated with the morals and customs of their hosts and barely able to speak their mother tongue. Their re-adaptation was not a success.

'During the night the news had spread, and early in the morning [23 January 1833] a fresh party arrived. Several of them had run so fast that their noses were bleeding, and their mouth frothed from the rapidity with which they talked, and with their naked bodies all bedaubed with black, white, and red, they looked like so many demoniacs who had been fighting.'

Journal of Researches

hair was tangled, their mouths frothed with excitement, and their expression was wild, startled, and distrustful. They possessed hardly arts, and like wild animals lived on what they could catch; their had no government, and were merciless to every one not of their one small tribe. He who has seen a savage in his native land will not feel much shame, if forced to acknowledge that the blood of some more humble creature flows in his veins.'

Commentators taking this text out of context have too often read into these lines, or those that correspond to them within the account of the voyage, some trace of personal 'racism' on Darwin's part, a racism that was later applied during the development of the theory of natural selection. This misinterpretation has been much more harmful than the traveller's words themselves, which must, of course, be read right to the end.

It is true that Darwin, as a naturalist, drew a link between human civilization and animal domestication, while already suggesting the crucial idea of an unbridgeable gap between their respective potentials. It is also undeniable that he deplored the

B elow, a member of the Tekeenica tribe (from which Jemmy Button had come), drawn by Conrad Martens. Left, the Beagle at Tierra del Fuego, watercolour by Conrad Martens, 1834.

Fuegians' nakedness, their lack of industry, their 'abject' attitudes and the use of an idiom which, 'following our conceptions' – this stipulation should be stressed – scarcely merited the term 'articulated language'. His notes dwell on the islanders' remarkable talent for imitation (of expressions, gestures and sounds), which he would later recognize as a general trait among 'savage' peoples, and which would form part of the behavioural evidence that would enable later evolutionary anthropologists to classify apes, savages, children and idiots all together, as being in various ways 'witnesses of the origin'. He also noted the resemblance between the vocal and gestural expressions of surprise among the Fuegians and the same expressions in an orang-utan at the zoo.

At the same time, however, he acknowledged the inability of any civilized European to repeat as perfectly as the Fuegians complete sentences in an unknown language, and he attributed their behaviour to their isolation and the harsh climatic conditions. Finally, and most remarkably, it is at this point in his narrative that he mentions for the first time the three Fuegians who had sailed with the *Beagle* from England, and who were being returned to their native land after three years away, converted to Christianity, speaking broken English and filled with the principles of 'civilization', as living proof of the ability of all human beings to develop their intellectual and moral faculties through education.

Racism in its most basic form is non-evolved (ethnocentric) behaviour, within which sympathy is limited to a restricted group. Among 'civilized' people, Darwin would, with strict logic, view it as a sort of throwback, a regression to a 'barbarous'

Canis antarcticus (below), native to the Falkland Islands, described in the *Zoology of the Voyage of H.M.S. Beagle* in 1839 as 'intermediate between the ordinary foxes and the wolves.'

It was quite melancholy leaving the three Fuegians with their savage countrymen; but it was great comfort that they had no personal fears. York, being a powerful, resolute man, was pretty sure to get on well, together with his wife Fuegia. Poor Jemmy looked rather disconsolate, and would then, I have little doubt, have been glad to have returned with us. His own brother had stolen many things from him; and as he remarked, "What fashion call that?" he abused his countrymen, "all bad men, no sabe (know) nothing", and though I never heard him swear before, "damned fools". Our three Fuegians, though they had been only three years with civilised men, would, I am sure, have been glad to have retained their new habits; but this was obviously impossible. I fear it is more than doubtful whether their visit will have been of any use to them.

Journal of Researches
[Left, the *Beagle* in the Murray Narrows, Tierra del Fuego in 1833, watercolour by Conrad Martens.]

human state; civilization is, in contrast, directly proportional to the extension of sympathy, so racism, like cruelty, is its very opposite. Darwin developed this theme in 1871.

On 6 February 1833, the *Beagle* picked up Father Matthews, who had been set down in a cove fifteen days earlier with the three Anglicized Fuegians in an attempt to set up a mission there, and who had been completely robbed by the natives. A few days later, the *Beagle* set sail for the Falkland Islands, claimed by the British a year earlier, where it stayed between 1 March and 6 April, before returning to Maldonado, and then Argentina, where it arrived on 3 August. It was now that Darwin's overland travels produced their most remarkable results.

The fossil mammals of Argentina

During an expedition heading for Buenos Aires, in the month of August 1833, Darwin unearthed,

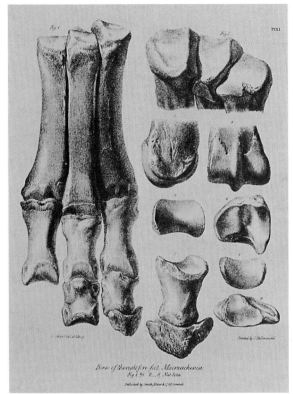

Bones of the right front feet, Macrauchenia
Fig 1 2_9 Nat Size
Published by Smith, Elder & Co. Cornhill

B ones of the right front paw (left) and skull – with a full set of teeth and normal wear of the incisors (above) – of *Macrauchenia patachonica*, 'a large extinct Mammal, classifiable in the Order of the Pachyderms, but with affinities with the Ruminants, and especially with the Camelids,' according to Richard Owen, who examined the fragments of the fossil mammals brought back by the *Beagle*. Its name means 'big neck'. The animal bore some resemblance to the llama, as well as the tapir, and today it is classed with the three-toed litopterns (an order of fossil ungulate mammals unique to South America). It has a nasal orifice on top of its skull, between the eye sockets, which has led some palaeontologists to suppose that it had a short trunk. The interest of *Macrauchenia* was partly due to the abundance of its fossilized remains.

in the layers of gravel and reddish mud of Punta Alta, a great many remains of fossil edentate mammals of the Quaternary era (*Megatherium*, *Megalonyx*, *Scelidotherium* and *Mylodon*, giant land sloths). He was also struck by the remarkable morphological resemblance between giant fossil armadillos and modern-day species. The presence nearby of shells that were very close to present-day types confirmed in his mind the validity of Lyell's ideas on the inferior longevity of mammalian species in relation to that of mollusc species.

Five months later (January 1834), and farther south, on the coast of Patagonia, he excavated from the red mud covering the gravel of the plain, 27 m (90 ft) above sea-level, half of a skeleton belonging to *Macrauchenia patachonica*, a hoofed quadruped the size of a camel, which was then classed among the pachyderms despite its elongated neck which resembled that of a giraffe. Since recent sea shells could be found in great number on two raised plains beyond the site, he concluded that *Macrauchenia* must have existed even more recently. This was the genesis of an important idea, anticipating the later development of the theory of descent, which explained the visible bonds of kinship between living species and fossil species of South American mammals: *Macrauchenia* is related to the guanaco, the toxodon to the capybara, the extinct edentates to modern sloths, anteaters and armadillos. Reflecting on the causes of the relatively recent extinction of the great animals that were found in a fossil state, Darwin examined and then rejected the hypothesis of a catastrophe which, bearing in mind the extent of the area concerned, would have shaken the globe more

Reconstructions of the skeletons of a mylodon (below left) and a megatherium (below right), both giant fossil edentates. All that was known of the mylodon (named *Mylodon darwinii* by Owen) at that time was a damaged lower jawbone around 44 cm (17 in) long and some teeth, while *Megatherium cuvieri* was represented by a skull fragment and some teeth.

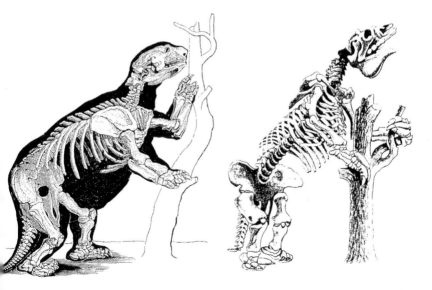

severely, and which was also ruled out by
observation of the gradual nature of geological
changes in the regions of La Plata and Patagonia.
The research also showed that extinction is
always preceded by a growing scarcity, the rarity
of the representatives being the evidence for
less favourable living conditions for the species.
This provides more evidence in favour of
gradualism, which Darwin always recognized
as a dominant feature of evolutionary processes:
transmutations occur continuously and by often
imperceptible degrees, following a rule of slow
progress and accumulation of slight variations
over the course of long periods of time.

Palaeontology and biogeography

The South American stage of the voyage was
exciting. In October 1833, Darwin studied the
geology of the Pampas, and discovered not only
fossil evidence of its recent rise above sea-level (in the
form of sea shells belonging to present-day species),
but also important data illustrating the processes
of biogeographic distribution: the replacement of
species, migration, and the role of geographical
barriers. This evidence was acquired while exploring
the region of Santa Fé Bajada, and secondary findings
later confirmed it: Darwin unearthed teeth belonging
to the toxodon and mastodon, as well as a tooth from

Present-day edentates,
from top: pangolin
(order Pholidota; Asia
and Africa); and three
species of armadillo
(order Xenarthra, which
also includes anteaters
and sloths; America),
from 19th-century
engravings. The third
order, Tubulidentata
(*Orycteropus*; Africa) is
not displayed here. Left,
a naturalized armadillo
(*Dasypus*) brought back
from Bahía Blanca in
Argentina by Darwin in
1833. Opposite, three
different species of
Argentine palm trees:
*Cocos yatai, Copernica
cerifera, Cocos
australis.*

a fossil horse common to both parts of the American continent (Owen's *Equus curvidens*) which had been supplanted by the descendants of the horses imported by the Spanish.

On the basis of this information, Darwin devised a likely scenario for biogeographical distribution and animal migrations. He suggested that America is divided along a line passing south of Mexico, at the point where the great plateau forms a natural obstacle to the migration of species. This line separates two faunas that are still distinct today, with the exception of a few species (puma, opossum, kinkajou, peccary) that have succeeded in crossing the barrier. The zone extending south of this line was later designated the 'neotropical' region. However, the study of fossil species brings to light a much greater homogeneity between these two territorial zones, clearly indicating that they were not differentiated until relatively recently, when the Mexican plateau was uplifted, or more probably, as Darwin thought, when the lands in the region of the Antilles subsided. Similarly, identical fossil remains of elephants, mastodons, horses and hollow-horned ruminants are found on either side of the Bering Strait, indicating that the north-western

part of North America and the north-eastern part of Asia were once joined together.

Exploring the Andes

There followed six months of travelling and exploration: the straits of Magellan, a second visit to Tierra del Fuego, Beauchene Island, the Falkland Islands, the ascent of the Santa Cruz river; Patagonia, where Darwin made some interesting geological observations and hypothesized about the glacial transport of erratic rocks; Cape Gregory, where FitzRoy dined with three Patagonian giants; Cape Turn, entry to the Pacific Ocean; the west coast of Tierra del Fuego, where Darwin considered the geological action of glaciers; and the Chilean island

On 16 April 1834, on its return from Tierra del Fuego and the Falkland Islands, the *Beagle* was hauled ashore on a bank at the mouth of the Santa Cruz river in Patagonia for some minor repairs (opposite). The next stop was Chile, where Darwin made his first great overland trip, in the foothills of the Cordilleras (above). On 14 August, he set out on horseback 'for the purpose of geologizing the basal parts of the Andes' in a 'grandiose and sublime' landscape.

of Chiloé. The *Beagle* reached Valparaiso, Chile, on 23 July 1834. Darwin made his first overland expedition to the foot of the Andes between 14 August and 27 September 1834, followed (after a second visit to Chiloé and a visit to the Chonos Islands) by a crossing of the Cordillera to Mendoza (18 March to 17 April 1835), and finally a third trip, to Coquimbo and Copiapó. Between the first two expeditions, in Valdivia on 20 February 1835, he experienced an earthquake whose effects he observed shortly afterwards on the island of Concepción. In the Andes, it was geology that once again occupied most of his attention. The young Darwin, filled with his readings of Lyell, was fascinated with the science because it was proof at the most basic level, within the very structure of landscapes, of the omnipotence of time.

'Who can avoid', he wrote in his *Journal*, 'admiring the wonderful force which has upheaved these

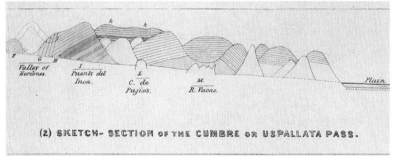

(2) SKETCH- SECTION of the CUMBRE or USPALLATA PASS.

mountains, and even more so the countless ages which it must have required, to have broken through, removed, and levelled the whole mass of them? It is well in this case to call to mind the vast shingle and sedimentary beds of Patagonia, which, if heaped on the Cordillera, would increase by so many thousand feet its height. When in that country, I wondered how any mountain-chain could have supplied such masses, and not have been utterly obliterated. We

Above, Darwin's sketch, published in 1846, of a section of the central chain of the Chilean Cordillera: 'The range resembled a great solid wall, surmounted here and there by a tower, and thus made a most complete barrier to the country.' (*Journal*, 17 August 1834).

'The earthquake commenced at half-past eleven in the forenoon. If it had happened in the middle of the night the greater number of the inhabitants (which in this one province amount to many thousands), instead of less than a hundred, must have perished.... In Concepción, each house, or row of houses, stood by itself, a heap or line of ruins; but in Talcuhano, owing the great wave, little more than one layer of bricks, tiles, and timber, with here and there part of a wall left standing, could be distinguished.... A bad earthquake at once destroys the oldest associations: the world, the very emblem of all that is solid, has moved beneath our feet like a crust over a fluid.'

Journal of Researches

L eft, the ruined cathedral of Concepción (Chile) after the earthquake of 1835. Below, Darwin's geologist's hammer.

must not now reverse the wonder, and doubt whether all-powerful time can grind down mountains – even the gigantic Cordillera – into gravel and mud.'

During the second expedition, from Santiago to Mendoza, Darwin studied the terraces of pebbles which extend on both sides of the great valleys, and confirmed his theory about the gradual elevation of the ground. Biogeography was far from forgotten, however; here for the first time he began to connect it with the study of the physical structure of these regions. He could not help but notice the considerable differences between the vegetation of the eastern valleys of the Cordillera and that of the Chilean valleys, despite their similar weather and soil conditions. The animals seemed to follow the same law of differentiation; Darwin recorded thirteen species of mice on the Atlantic side, but only five on the Pacific side, with no resemblance between them. From this he concluded that the Andes have formed a virtually impenetrable geographical barrier since the appearance of modern animal species. However, most of the plant and animal species of the eastern plains are very close to those which inhabit Patagonia.

The *Beagle* left Chile for Peru, where it stayed for two months before setting out to sea once again.

The Galapagos islands

From 15 September to 20 October 1835 Darwin visited the Galapagos archipelago, located in the Pacific Ocean on the Equator, about 1000 km (600 miles) from the coast of Ecuador. He studied the flora and fauna of these volcanic islands, particularly the reptiles and birds. The geologically recent nature of these lands presented him with a mystery: they were populated by creatures which, he noticed, were distinctly native in character, yet showed kinship with their probable South American rootstock. Darwin dissected and brought back specimens of native lizards (in particular two species of iguana, one marine and the other land-based, which he noted were incredibly profuse), and he later reproached himself for not having paid more attention to the giant tortoises (*Testudo nigra*) which at the time were in their egg-laying season. All these animals, from one island to the next in the archipelago, displayed differences which seemed to be at the variety level. This at least is what Darwin believed as regards a distinctive group of small land birds. The size of sparrows or finches, this group are, with one small exception, unique to the archipelago and after Darwin's return, the ornithologist John

Cactornis scandens (below) is one of the Galapagos finches. The black male has a tapered beak. It might be guessed from the shape of its beak that it does not feed by crushing, like *Geospiza magnirostris*, but in a manner similar to *Certhidea olivacea*, which is adapted to catch insects that live inside plants. In fact, *Cactornis* strikes the outside of a branch with its beak, then brings its head close to listen for the movements of the disturbed larvae, which it then extracts, using a cactus spine which it breaks off as a tool.

1. Geospiza magnirostris.
3. Geospiza parvula.

2. Geospiza fortis.
4. Certhidea olivacea.

FINCHES FROM GALAPAGOS ARCHIPELAGO.

Left, the heads of four Galapagos finches. Between *Geospiza magnirostris*, which has the strongest beak (see the pair below) and *Certhidea olivacea*, the sub-group with the thinnest beak, there are in fact even more species with beaks of gradually decreasing size. Darwin's paper, which owed a great deal to the identifications of John Gould, was delivered on 10 May 1837 to the Zoological Society, and made a major contribution to 'transformist' zoology by demonstrating the possibility of speciation (formation of species), caused by adaptation and insular isolation, from a single continental rootstock.

Gould identified as them distinct. Gould classified them into 13 species, divided into 4 sub-groups: *Geospiza* (8 species), *Camarhynchus* (2 species), *Cactornis* (2 species) and *Certhidea* (1 species). In general, the males are intensely black, and the females rather brown, and their basic shape led to their being identified as finches. Darwin concluded that, from a single rootstock species of continental origin, their isolation on distant islands had led them to develop marked variations, especially in the shape and size of their beak, that were probably linked to differences in lifestyle and feeding habits. It was in 1837, after Gould's assessment, that he recognized that these varieties were actually separate species. This confirmed an intuition that he would later develop in *Origin of Species*, and in his book on variation in 1868: varieties are nascent species. Although he did not realize it at the time, he had just observed a case of insular speciation.

Coral reefs

On 9 November 1835, the *Beagle*, en route to Tahiti, entered the Low (or Tuamotu) archipelago, where Darwin had his first chance to observe

Opposite, the giant tortoise of the Galapagos Islands, *Testudo elephantopus*. These enormous tortoises were named *Testudo nigra* by Darwin, who climbed on their backs, dissected at least one (he says that he found a caterpillar in the stomach of one of them), and was amazed by their ability to find water sources, trampling wide paths behind them. Their internal water reserves were sometimes used by thirsty travellers. Above left, the land iguana (*Amblyrhynchus demarlii*) in its natural habitat. Its tail is round and its feet are not webbed like those of its aquatic 'cousin'. Darwin dissected and studied both species. The land *Amblyrhynchus* lives in burrows and feeds mostly on cactus and berries, while the marine species eats only sea plants. Below left, the marine iguana *Amblyrhynchus cristatus*, whose compulsion to return to the shore was observed by Darwin. It has a laterally flattened tail, and partially webbed feet.

atolls; coral rings at water level, enclosing a lagoon. After several short stopovers (Tahiti, New Zealand, Australia, Tasmania), the *Beagle* gave Darwin another important opportunity to use his scientific acumen when, between 1 and 12 April 1836, he studied the structure of an atoll and examined the living coral in the Keeling or Cocos Islands in the Indian Ocean, about 1000 km (600 miles) from the coast of Sumatra. His study, which linked geological and biological phenomena, laid the foundations for all modern theories on the formation of coral reefs.

Mauritius, the Cape of Good Hope, the islands of Saint Helena and Ascension, and then Bahia once again, Pernambuco (Brazil), Cape Verde, and finally the Azores were the *Beagle*'s last stops on its journey home to England. On 2 October 1836, the ship arrived in the port of Falmouth; its voyage had lasted almost five years. Throughout the trip, Darwin had written a log in which,

These aerial photographs of the Polynesian islands of Moorea, Bora Bora and Tetiaroa (above) admirably illustrate the stages of formation of an atoll as reconstructed by Darwin in an engraving (below) from *The Structure and Distribution of Coral Reefs*, 1842.

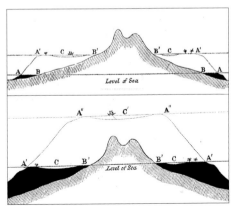

day by day, he had recorded the main events of shipboard life, his explorations and scientific observations, his thoughts and emotions, such as those he felt on seeing the lushness of the primeval rainforest or the immensity of the plains of Patagonia. He had also used the stopovers to send letters to his family, friends and former teachers. Henslow in particular was a favoured correspondent who regularly received mail and specimens.

The results of the voyage

On his return, Darwin was no longer an unknown. His correspondence with Henslow and Sedgwick meant that he was now recognized as a promising naturalist and geologist. Having quickly renewed his many friendships in Cambridge, he set to work and, simultaneously, prepared to publish his journal, finished several monographs, and busied himself with the publication of the scientific results of his voyage.

The specimens brought back on board the *Beagle* were entrusted to different specialists: the fossil mammals to Richard Owen, the modern mammals to George Robert Waterhouse, the birds to John Gould, the fish to Leonard Jenyns, and the reptiles to Thomas Bell. Queen Victoria came to the throne on

The base of an atoll or barrier reef is made of dead coral, and living organisms only occupy the last few dozen metres. Guided by Lyell's theory of compensatory movements of the earth's crust, Darwin suggested the following process: an emerged volcanic island is surrounded and fringed by a reef (A-B, B-A). The island slowly sinks. Coral at the base is deprived of light, and dies. New coral grows at the top, compensating for the subsidence and increasing the diameter of the reef. If these movements balance each other out, the reef becomes a barrier (A'B'-B'A') separated from the central peak by a channel lagoon. When the summit of the island is submerged, what remains is an atoll (A''), a ring of coral with an enclosed lagoon.

20 June 1837, and that same year saw Darwin's election to the Geological Society; he became its secretary on 16 February 1838. The scientific volumes that were a direct result of the voyage were published over a span of more than eight years: the *Journal of Researches* in 1839, the *Zoology* from 1838 to 1843, the *Geology* from 1842 to 1846.

The impact of the voyage was enormous and wide-ranging, setting in place elements of observation and induction which future theories would eventually link together. On his return, Darwin realized the usefulness of uniformitarian geology in explaining phenomena such as the formation of volcanic islands, the construction of coral reefs, the uplift of the American continent, and the compensatory alternation between movements of elevation and subsidence of the ground. He also learned the importance of constant methodological attention to the power of time and to repeated and cumulative actions on a small scale.

DARWIN

THE

ZOOLOGY

OF

THE VOYAGE OF H.M.S. BEAGLE,

UNDER THE COMMAND OF CAPTAIN FITZROY, R.N.,

DURING THE YEARS
1832 TO 1836.

Wherever he went, he observed the determining influence that living conditions and climate had on organic life. He carefully noted the geographic distribution of animals, and recognized the isolating function of land or sea barriers. He highlighted the continuing similarities between living and extinct animal species within a single geographical region. He noticed the transforming effects of domestication on animals, and observed the effects of acclimatization on imported plants. He noticed the modifying and

Right, this scorpion fish from the Galapagos (*Scorpaena histrio*), has spines of unequal length covering the top of its back, a small triangular spine on each of its nasal bones, and spines above its eyes. More spines, not visible here, cover other parts of the head. Left, title page from the first edition of *Zoology*.

sometimes destructive influence of human action on nature and its primitive equilibria. He conjectured, especially in South America, as to the reasons why species become extinct, and speculated on the importance of food resources and the need for factors of demographic regulation. He recognized that there are mechanisms of interaction between a species and its environment, and between one species and another. He put forward hypotheses on migrations and the modes of transport of individuals and seeds. In the Galapagos, he noted, although he had no conclusive proof, the existence of a process of insular speciation that was clearly adaptive. He observed the extreme diversity of human beliefs, customs and rites, experienced the poverty of the condition of savage man, but also

Corals are colonies of invertebrates. A root polyp fixes itself to a spot and, through a process of budding, produces other polyps that are linked by their tissues and their chalky skeletons (polyparies). Biologically related to single-celled green algae, polyps require light, and die below a depth of 80 m (260 ft). They bloom at a temperature between 25°C and 27°C, and at a depth of 30 to 35 m (100 to 115 ft). Left, *Anthozoa octocorrallia alcyonum*, although not a reef-building species, gives a rare display of its open polyps.

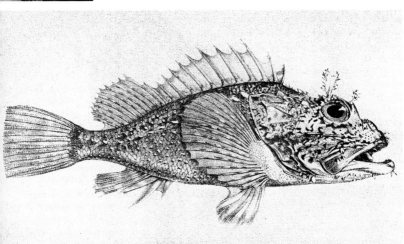

tested his capacity for civilization, and fiercely criticized the the practice of slavery, whose ignominy was masked by the demand for a 'right' naturally possessed by the 'superior' over the 'inferior'.

Gould's appraisal of the bird findings, on which Darwin based a paper he delivered to the Zoological Society on 10 May 1837, immediately showed that the different finches of the Galapagos were in fact species, and not simply varieties, favouring a 'transformist' interpretation. In July Darwin began to make his first notes on the 'transmutation' of species, as well as embarking on a vast programme of reading in all the natural and human sciences. He also continued work on his study of coral reefs, which in 1842 formed the first volume of the *Geology of the Voyage of H.M.S. Beagle*.

Malthus and population growth

Between 27 September and 4 October 1838, Darwin read a book that was to have a significant influence on his work: *Essay on the Principle of Population* by Thomas Robert Malthus, an English economist and sociologist who was also a parson. Speculating on the conditions for social progress, he had

Thomas Robert Malthus (1766–1834; left) applied the model of competition and elimination to human populations. Darwin first applied it only to plant and animal populations (*Origin of Species*, chapter III, and *Variation*, Introduction). and later extended it to humans (*Descent of Man*, chapter XXI: 'Man tends to increase at a greater rate than his means of subsistence; consequently he is occasionally subjected to a severe struggle for existence, and Natural Selection will have effected whatever lies within its scope.').

discovered and published in 1798 what he called the 'principle of population': the human population increases far more quickly (geometric progression) than food can be produced (arithmetic progression). He believed that this periodically caused huge and destructive catastrophes such as wars, famines and epidemics, and that these could be avoided by limiting birth rates, especially among the lower classes.

However, he never subscribed to Malthus's suggestion that births in human societies should be limited, nor to his theological providentialism (hardship is instituted by God to provoke effort and to make man desire and deserve the next world). Nor did Darwin believe in the notion of society being frozen in a state of fixed equilibrium that excludes all prospect of progress, and even less in the refusal of assistance to the poor. Darwin explains his position openly in *The Descent of Man* (chapter

But I have too deeply enjoyed the voyage not to recommend to any naturalist to take all chances, and to start on travels by land if possible, if otherwise on a long voyage.

Although in 1838, Malthus provided Darwin with a basic scheme that helped to shape the theory of natural selection, Darwin did not support his social philosophy or the bullying tactics he suggested.

When Darwin wrote the first two outlines of his theory, in 1842 and 1844, he already had at his disposal the principal phenomena which had to be explained (biogeographic distribution, adaptive variations, extinctions, reproduction rates, demographic equilibria, the effects of domestication) as well as the principle by which they were connected. He knew that a process of competition and elimination had to exist, which used struggle and sexual selection as a means of regulation.

XXI): 'Hence our natural rate of increase, though leading to many and obvious evils, must not be greatly diminished by any means.' This allows for free competition and is also a vote in favour of civilization, where the measure of progress is the capacity for the reparation of suffering. Above, Darwin's desk at Down House, with some of his notebooks and equipment.

'**R**eflect on the enormous multiplying power inherent and annually in action in all animals; reflect on the countless seeds scattered by a hundred ingenious contrivances, year after year, over the whole face of the land; and yet we have every reason to suppose that the average percentage of every one of the inhabitants of a country will ordinarily remain constant.'

Essay of 1844

CHAPTER 3

THE THEORY OF NATURAL SELECTION

Opposite, Charles Darwin and his eldest son, William Erasmus (Doddy), in the summer of 1842. Darwin's daily observations of his baby son's reactions and expressions later became his *Biographical Sketch of an Infant*, published in 1877. Right, a 1837 sketch of the branching formation of genera.

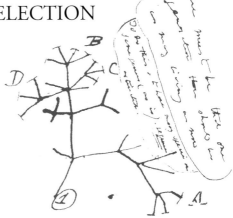

Everything begins with variation. Individual lifeforms vary, and so do their habits, behaviour and instincts, and Darwin had long been aware of this fact. He wrote that variation is 'random', by which he meant that its causes were as yet unknown, and to explain it he invoked the (still mysterious) laws of heredity.

The best observatory and laboratory of variation is the horticulturalist's garden or the stockbreeder's enclosure: here, variant forms can be seen as soon as they emerge, then sorted and either discarded or, alternatively, kept and allowed to reproduce if they have a particular advantage to offer their human breeders. This is the process of artificial selection, and always implies the elimination of individuals which do not conform to a norm of excellence that is extrinsic to the organisms themselves.

Darwin believed that for any organism, a change in conditions is a trigger for change. The survival of such modifications is subject to new adaptive constraints, which favour the appearance of variations different from those prevalent up to that point. Since domestication itself constituted a considerable change in conditions, Darwin studied the phenomenon and tried to verify that variability increases perceptibly, both in scale and rapidity, through the effects of cultivation, stockrearing and even of fashion in the case of flowers or pets. In 1839, after writing his first notebooks on the 'transmutation' of species (1837) and reading Malthus (1838), Darwin sent out a questionnaire to English stockbreeders, to collect information relevant to his theories in this field.

During the process of domestication,

•The initial variation on which man works, and without which he can do nothing, is caused by slight changes in the conditions of life, which must often have occurred under nature. Man, therefore, may be said to have been trying an experiment on a gigantic scale; and it is an experiment which nature during the long lapse of time has incessantly tried.•
The Variation of Animals and Plants under Domestication, Introduction

Below, a powerful Normandy horse, built for harness work. Opposite, a Barbary horse, originating in North Africa, built for battle and racing.

man selects variations which are potentially useful. However, the organisms selected in this way nonetheless remain natural organisms. In fact it is nature, and nature alone, which provides the variations that man keeps and maintains, using them to shape domestic breeds to human advantage. From this, Darwin deduced that nature operates in an analogous way, only over much longer periods. The forces which regulate this process work to the advantage of the organisms themselves. Within the struggle for life – including the competition produced by the spontaneous growth of specific populations (the Malthus model) – the suitability or unsuitability of particular characteristics to an environment and limitations on available resources ensure the automatic sorting of individuals that carry useful variations which will improve the adaptation, and the correlative elimination of individuals which that lack this variation. This is natural selection.

Critical years

In 1839, having settled in London, Darwin was elected a fellow of the Royal Society. He saw a great deal of Lyell, with whom he debated

This cartoon sums up natural selection, illustrating the success of an advantageous variation (an increase in height, as in the classic example of the giraffe) within a specific environment (a dry climate, where the only fresh vegetation available is at the top of trees). Darwin was of course aware that a long process of accumulation of small variations is required, each conferring a slight advantage, to produce an adaptation.

**First input
(facts 1 and 2)**

1. Living organisms, both domesticated and wild, display variations.
2. Therefore there is a natural variability of organisms.
3. Man can select from these variations to his own advantage by selecting individuals and guiding their reproduction.
4. Therefore there is a natural 'selectability' of organisms.

Question: Natural selection has been inferred as a possibility. If this is the case, how does it operate?

**Second input
(facts 3 and 4)**

5. The birth rates of different species are evaluated and their ability to reproduce indefinitely is established.
6. From this can be deduced a mathematical law: all space will become rapidly overpopulated by the members of any species that can reproduce without obstacle.
7. However, this state of absolute saturation never occurs. Instead, what is observed is the balanced coexistence of many species within the same territory.
8. From the opposition between points 6 and 7 (deduction 3 and fact 4)

1

Fact 1
**Variation
(natural or
domestic)**

3

Fact 2
**Artificial selection
(horticulture,
stockbreeding)**

2

Induction 1
**Natural capacity
for variation
(variability)**

4

Induction 2
**Natural capacity
for being selected
(selectability)**

*Question
Does selection of variations
occur in nature?*

10

Hypothesis
**Selection of
advantageous
variations**

*Question
What determines
best adaptation*

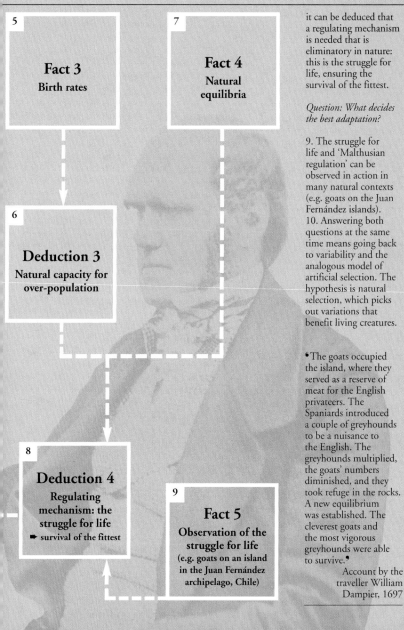

5

Fact 3
Birth rates

7

Fact 4
Natural equilibria

6

Deduction 3
Natural capacity for over-population

8

Deduction 4
Regulating mechanism: the struggle for life
➤ survival of the fittest

9

Fact 5
Observation of the struggle for life
(e.g. goats on an island in the Juan Fernández archipelago, Chile)

it can be deduced that a regulating mechanism is needed that is eliminatory in nature: this is the struggle for life, ensuring the survival of the fittest.

Question: What decides the best adaptation?

9. The struggle for life and 'Malthusian regulation' can be observed in action in many natural contexts (e.g. goats on the Juan Fernández islands). 10. Answering both questions at the same time means going back to variability and the analogous model of artificial selection. The hypothesis is natural selection, which picks out variations that benefit living creatures.

•The goats occupied the island, where they served as a reserve of meat for the English privateers. The Spaniards introduced a couple of greyhounds to be a nuisance to the English. The greyhounds multiplied, the goats' numbers diminished, and they took refuge in the rocks. A new equilibrium was established. The cleverest goats and the most vigorous greyhounds were able to survive.•

Account by the traveller William Dampier, 1697

questions of geology. He married his first cousin Emma Wedgwood (daughter of Josiah II), and was present at the birth of their first child, William Erasmus; he went on to meticulously record all the child's expressions and emotions for almost two years. This was also the year of the publication and success of his *Journal of Researches*, his first meeting with his future friend, the botanist Joseph Dalton Hooker, who was embarking for the Antarctic; and the distribution of his questionnaire to stockbreeders. It also heralded the worsening of a chronic illness, probably caused by the bite of a poisonous insect in South America and diagnosed more recently as Chagas' disease, which would burden him till the end of his life.

While he was pursuing his theories on species, he expanded and diversified his reading matter. He read Lamarck and his own grandfather, Erasmus Darwin, Linnaeus on the economy of nature, Johannes Müller on physiology, Charles Bell on expressions, William Buckland on geology, and also took an interest in reproduction, instincts, philosophy and ethics.

The Royal Society (above) began as an informal club in 1645. It found a permanent home in London in 1659, and was given a charter and its current name by Charles II in 1662, as being of benefit to the public. Its purpose was to foster research and scientific exchanges of all kinds; discussion of political or religious subjects was expressly forbidden. Erasmus Darwin was elected a fellow in 1761, Josiah Wedgwood in 1783, Robert Darwin in 1788. Three of Charles's five sons (George, Francis and Horace) later became fellows.

His first daughter, Anne Elizabeth, was born in 1841. In 1842 he published his book on coral reefs, and the Darwins settled at Downe (Kent), south-east of London, in a country dwelling called Down House. A little girl, Mary Eleanor, was born and died between September and October of that year. Darwin pencilled down, in note form, the first outline of his theory. Another girl, Henrietta Emma, was born in 1843, at the time when he began his friendship with Hooker, his notes on variation, and his work on double flowers. In 1844, having been elected vice-president of the Geological Society, he wrote an essay of about 200 pages (the second outline), while Robert Chambers, a publisher and amateur geologist, anonymously published *Vestiges of the Natural History of Creation*, a 'transformist' work which was widely criticized by the scientific community, which gave Darwin cause to be prudent. George Howard was born in 1845, followed by Elizabeth (1847), the future editor of the family's correspondence, then Francis (1848), Leonard (1850) and Horace (1851). By 1846, the entire *Geology of the Voyage of H.M.S. Beagle* had been published, and Darwin devoted the next eight years to a four-part monograph on some common but little studied crustaceans, the cirripedes. The work was published between 1851 (the year when his daughter Anne Elizabeth died, at the age of ten) and 1854.

The cirripedes

This group of marine invertebrates, which

Emma Darwin (1808–96; pastel drawing from 1840). Her devotion to Charles was matched in him by deep feelings of gratitude and respect.

Joseph Dalton Hooker (1817–1911; left) was a botanist like his father, William Jackson Hooker (1785–1865), whom he succeeded as director of the Royal Botanical Gardens at Kew. He made two voyages: one to the Antarctic (1839–43), and the other to the Himalayas (1847–50). One of the first to learn of Darwin's theory, he gave it his support in October 1859, just before the publication of *Origin of Species*. Darwin held him in high esteem, and considered him the only man who had always been on his side.

Down House, seen from the garden in this anonymous etching (left), was the home of the Darwin family. Opposite below, the photograph shows, from left to right: Leonard, Henrietta, Horace, Emma, Elizabeth, Francis and a visitor. Below, Darwin on the back of his old horse, Tommy, in the early 1870s.

includes the barnacles, belong to the sub-class of fixed crustaceans, but previously had been wrongly believed to be molluscs (Cuvier was one of the last to make this mistake) and had long misled classifiers. Darwin was greatly interested in classification difficulties, because they bear witness to the fragility and contradictions of old creationist and typological conceptions of nature. Moreover, organisms which take on different forms at different stages of their lives

are testament to the fact that change, rather than 'fixity', is the law of the organic world. Because of this, Darwin was fascinated by metamorphoses, the alternation of generations, the conversion of some organisms from a mobile form to a fixed form at different stages in their development, and the movement of plants. The analogies or affinities between zoological groups (similar metamorphoses in insects and crustaceans, for example) and even between kingdoms (the 'animal' traits of carnivorous plants) strongly supported the idea of a common ancestry for all organisms, and a genealogical representation of living things. Darwin, as his correspondence with Hooker of September 1845 shows, was also anxious

Above left, *Pollicipes* or goose barnacles, pedunculated cirripedes (*Lepadidae*). Below, outer and inner views of another pedunculated cirripede, *Anelasma* (Darwin, 1851).

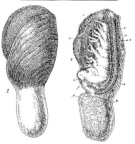

Acorn barnacles (left, photograph and below, sketches of an exterior view and cross-section, 1854), are stalkless (sessile), and stick directly to rocks, shells and boats. They also possess an operculum (closing plate). Cirripede larvae are free-swimming at first, then fix themselves by the antennules on their head end. The adults are fixed, and feed by making a water current with their cirri, whips covered with bristles situated near their mouth (these are unfurled in the

to be considered as more than just a lucky traveller: the Cirripedia books were his chance to be recognized as a real naturalist. Darwin confirmed that these creatures are in fact crustaceans (a fact first established through John Vaughan Thompson's study of their larvae in 1835) and produced a classification. His monograph was a great success, and was particularly admired by Richard Owen.

photograph). They are mostly hermaphrodites, but some species have dwarf or 'complemental' males which are extremely regressed, reduced to their reproductive organs; these fix themselves inside fully grown individuals (females or hermaphrodites).

The fragility of fatherhood

By the mid-1850s, Charles and Emma had completed their family. Their last child, Charles Waring, lived for only two years (1856–58), dying in the year that the theory of descent first appeared. Over a period of almost twenty years, ten children had been born, and three of them had died. Darwin's relentless promotion of the usefulness of studies into consanguinity might partly be explained by his ever-present awareness that he himself had married his first cousin. He loved Emma tenderly and devotedly, with mutual esteem and shared sorrow.

These years saw Darwin perfecting his fundamental theories on the geographical distribution of organisms, and also saw the appearance of a possible claim to co-authorship of the theory of descent with modification through natural selection from the young Alfred Russel Wallace.

A respectful rival

Alfred Russel Wallace was fourteen years younger than his rival. A naturalist, like Darwin, he had travelled extensively, visiting the Amazon basin with Henry Walter Bates (1848–52) and the Malaysian archipelago (1854–62). Like Darwin, he had studied specimens, visited islands and brought back collections. Like Darwin, he had published an account of his travels. Like Darwin, he had studied the new geology (Lyell), biogeography (Humboldt), the laws of population (Malthus) and classification. And like Darwin, he had come to the conclusion that living species are transformed, beyond simple variations, by the pressures of selection.

In September 1855, Wallace published an article 'On the Law which has Regulated the Introduction of New Species'. At the beginning of 1856, Lyell grew worried that Darwin might be in danger of being deprived of his due credit as a pioneer, and urged him to publish his theory. Darwin then began to write what became *Origin of Species*.

On 18 June 1858 Darwin received a manuscript called 'On the Tendency of Varieties to Depart Indefinitely from the Original Type', in which Wallace linked evolution to the mechanism of selection. Supported by his friends Lyell, Hooker and Huxley, Darwin accepted that the former should organize a presentation of his theory, including the manuscript by Wallace (who was in Malaysia) to the Linnean Society on 1 July 1858.

The great mutual respect of the two men and their exemplary regard for fair play won the day over any temptation to fight. Wallace, the potential rival, became a supporter and friend of Darwin. However, despite his remarkable intelligence, to Darwin's dismay he was shortly afterwards attracted by spiritualism, and once again gave an ambiguous place to the idea of a creator at the heart of his concept of human evolution.

Alfred Russel Wallace (1823–1913; opposite) was only 35 at the time of the Linnean Society presentation; Darwin was then almost 50. Darwin's precedence goes back to 1837 (the date of his notebook on transmutation), 1842 (first outline) and 1844 (second outline) and is therefore indisputable. Darwin had great respect for Wallace's skill as a naturalist, particularly on questions of biogeography and entomology, and also expressed admiration – together with some criticism – for Wallace's essay of 1864 on 'The Origin of Human Races and the Antiquity of Man Deduced from the Theory of "Natural Selection"'. However, he strongly disapproved of Wallace's penchant for the supernatural, and his belief, made explicit in 1869, in a transcendent evolution for man which reintegrated him into the plan of Providence. Man became, in a way, God's domestic animal, through an analogy between the intelligence of the stockbreeder improving his horses and the intelligence of a supreme being improving humankind towards a nobler end. Left, Darwin (seated, holding paper), Lyell (standing) and Hooker (seated), in a painting by an anonymous artist.

74

'Thus, from the war of nature, from famine and death, the most exalted object which we are capable of conceiving, namely, the production of the higher animals, directly follows.'

On the Origin of Species, 1859

CHAPTER 4

A RESOUNDING TRIUMPH

The first edition of *Origin of Species*, which applies the theory of natural selection to the plant and animal kingdoms, was published on 24 November 1859. However, it was not until 1871, the date of this jolly caricature in *Vanity Fair* (opposite), that Darwin's theory was extended to *The Descent of Man*.

THE ORIGIN OF SPECIES
BY MEANS OF NATURAL SELECTION,

The 1250 copies printed of the first edition of *Origin of Species* sold out on 24 November 1859, the first day of its publication by John Murray. On 26 November, Huxley published a eulogistic review in the *Times*. Six weeks later, on 7 January 1860, a new edition of 3000 copies appeared; Darwin made a few minor changes, and added a final remark including a mention of the 'Creator' in order to forestall the reactions of religious minds. The book had a stormy reception, causing conflict between supporters and opponents of the new theory, and, in broader terms, between scientific progressives and conservatives.

On the Origin of Species by means of natural selection, or the preservation of favoured races in the struggle for life is generally considered to be Darwin's most important work, because it advances the theory of descent by means of natural selection and its 'proofs' in nature: species are descended from one another by means of selected and transmitted adaptive modifications. Many ancestor species have become extinct, hence the absence of many 'intermediary forms'. However, the study of biogeographic distribution, rudimentary organs, the development of embryos, domestic cross-breeding, and classifications provides confirmation of the common ancestry of all living things, the divergence of characteristics, and the theory that varieties are nascent species.

Darwin remained deliberately silent about man in this first great treatise, with the exception of a single

Origin of Species was reprinted six times during Darwin's lifetime, in 1859, 1860, 1861, 1866, 1869 and 1872 (the 6th edition, whose final reprint in 1876 is considered definitive). Every edition after 1859 was expanded, revised and corrected by Darwin, who spent sixteen years trying to improve his text and answer to various criticisms. Below, the title from the manuscript of the future *Origin*, at that time a simple abstract on natural selection, begun in 1858.

prediction: 'Psychology will be based on a new foundation, that of the necessary acquirement of each mental power and capacity by gradation. Light will be thrown on the origin of man and his history.' He did not openly address anthropological questions until more than eleven years later, in *The Descent of Man*. Unfortunately, during this period of silence, others such as Spencer and Galton spoke in his place, and often in his name.

A sociological distortion

An English engineer of the Industrial Age, Herbert Spencer was the inventor of 'philosophical evolutionism', which he developed into a 'synthetic system' in 1860, and which spread around the world during the last third of the 19th century. He was a Lamarckian, and recognized the importance of Darwin's work, but this did not influence his grasp of biological phenomena, although it did serve him well in sociology. His 'system' was built around a theoretical core called the 'law of evolution' (the gradual passage, through integration and differentiation, from a structureless homogeneity to a structured heterogeneity), which he took from embryology and physics and then extended to all phenomena. He was also the father of the improperly named 'Social Darwinism', a crude application of the principle of selection and

Thomas Henry Huxley (1825–95; below), 'Darwin's bulldog', became a supporter in 1860, and fought against Owen and conservatism in science. This talented anatomist nonetheless had some reservations about natural selection, favouring evolution by sudden leaps, rather than the accumulation of slight variations. His *Evidence as to Man's Place in Nature* (1863) outlined the anatomical links between humans and the great apes.

elimination to human social groups, viewed as living entities. He personified a social philosophy that was ultra-liberal, individualist, and anti-state, strongly in favour of competitive conditions and fiercely hostile to the concept of aid for the poor. Darwin's *Autobiography* (1876) reveals a lack of taste for the man and a lack of interest in his ideas.

Francis Galton, an embarrassing cousin

An anthropologist and statistician, pioneer of biometrics and devotee of the study of hereditary traits, this young cousin of Darwin (he was proud to share Erasmus as a grandfather) represents the second distortion inflicted on the theory of natural selection during its author's silence on the subject of mankind. An admirer of Darwin, in 1865 he used elements of the theory of natural selection as the basis of what later came to be called eugenics: since natural selection ensured the diversity of species and the promotion of advantageous variations within the the living world, then the same thing should occur in society with regard to intellectual characteristics. Civilization hampers the free play of natural selection by protecting the survival and reproduction of 'mediocre' beings and causing a process of degeneration within the social group; this, he claimed, should be combatted through artificial selection, which

Francis Galton (1822–1911; left). Throughout his life, he nursed a passion for statistical studies of the lineages of illustrious thinkers. Above left, Herbert Spencer (1820–1903).

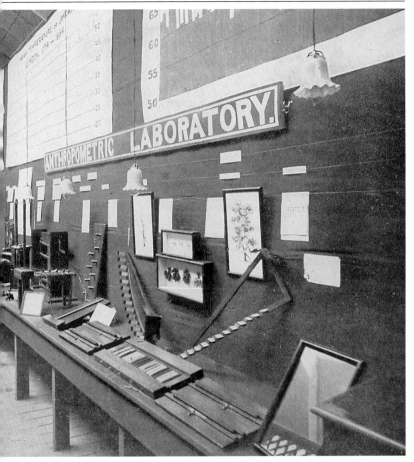

would compensate for the deficit and alleviate the burden on society. Although he respected Galton's work on statistics and took an interest in his 1869 book *Hereditary Genius*, Darwin rejected this misinterpretation of his theory in 1871.

The Battle of Oxford

In the early summer of 1860 there was held in Oxford an annual session of the British Association

The Anthropometric Laboratory (above), created by Galton. It opened to the public in 1884 as part of the International Health Exhibition at the Science Museum, London, and operated for five years. Inside it, families underwent a series of tests and measurements.

for the Advancement of Science which later became legendary. *Origin of Species* was already in its second edition, and Darwin had once again fallen ill and was undergoing treatment.

On Saturday 30 June, the bishop of Oxford, Samuel Wilberforce, drawing on his knowledge of natural theology and the active support of Richard Owen, the powerful head of the Department of Natural History at the British Museum, spoke before a crowd of almost a thousand people, and launched a harsh attack on Darwin's theory.

Using all the crowd-pleasing techniques of the orator's art with a crude opportunism, the bishop addressed T. H. Huxley, who was present along with other Darwinians such as Joseph Dalton Hooker and John Lubbock, and asked him if he preferred to be descended from a monkey through his grandmother or his grandfather. Huxley then gave a brilliant reply (but without, it seems, being heard by all of the assembled crowd). He explained the superiority of the theory of natural selection over any other hypothesis on the origin of species, stigmatized the scientific incompetence of the speaker, and ended by declaring to the bishop that he would rather be the grandson of an ape than related to a man who misused his mind on questions of such importance. Lady Jane Brewster, wife of David Brewster, the Scottish inventor of the kaleidoscope, fainted in the crowd.

In the midst of the resulting commotion, Hooker obtained permission from the president Henslow (his father-in-law and Darwin's old master and friend, although little inclined to support 'transformism') to address the assembly himself. His speech, brief and uncompromising,

Son of a statesman, Samuel Wilberforce (1805–73; below left) was bishop of Oxford. In 1860 he published a polemical article criticizing Darwin (who saw Owen's hand in it) in the *Quarterly Review*.

was in strong contrast with his very mild nature. He explained that the bishop could not really have read Darwin's book, and demonstrated that he did not know the first thing about botanical science. He received an ovation, and, according to the triumphant account that Hooker himself gave two days later in a letter to Darwin, the session closed with the bishop reduced to silence. Robert FitzRoy, former captain of the *Beagle*, was present at the meeting in his capacity as a meteorologist, but for his part could do no more than clutch his Bible and regret the publication of his former friend's book.

Through a combination of the many and diverse personal accounts by participants and journalists, the Oxford episode has become a legend, the tale of a decisive confrontation between science and religion. In fact, a more dispassionate view would take it as an important episode in science's emancipation from religious obligations of word and thought and their many metaphysical precepts; part of an unfinished battle in which science's only error was to believe its victory definitive.

Variation and heredity

Between 1861 to 1868, Darwin worked on several new editions of *Origin*, which from this point on

Richard Owen (1804–92; above), the anatomist who had reigned over the British Museum since 1856, promoted the doctrine of 'idealistic morphology', based on the notion of the archetype: the skeleton of each vertebrate is derived from an ideal vertebrate archetype and altered through secondary modifying factors, which God uses to create diversity among living things. Owen therefore claimed to have been the first to propose an evolutionary view of nature, which aroused Darwin's ire. He was fiercely opposed by Huxley (left) on the relationship of man to the great apes. Owen's views represented a compromise between French anatomy, the neo-Platonism of German Romanticism, and basic creationism.

included a Historical Sketch containing a meticulously detailed list of his 'transformist' predecessors. At the same time, he was also putting together his longest work of synthesis, *The Variation of Animals and Plants Under Domestication*, a development of the first two chapters of *Origin of Species*, and an enormous compilation of facts stemming from personal observations and contributions from many of his correspondents. This book examined the role of the direct influence of climatic and nutritional factors, the effects of the use and non-use of organs, the laws of heredity, the effects of cross-breeding on sterility in the domestic environment, the harmful consequences of repeated inbreeding, methods of selection, and the transmutation of domestic breeds. Among hundreds of other sources, Darwin used his own experience as a pigeon breeder. Nature provides man with variations whose source he does not know, and which he cannot therefore attribute to 'chance'. Human intervention consists solely of choosing to keep a variation if it offers a potential advantage, and increasing it through methodical selection and exclusive and prolonged cross-breeding of individuals that carry that trait. It is through this process of slow and directed accumulation that the domestic breeds are shaped and stabilized.

The example of the Galapagos finches had showed Darwin that under the pressure of different environments, nature alone could produce new species from a single original rootstock species. This would seem to confirm that, away from any artificial breeding process, advantageous variation represents the key to evolutionary divergence, and is the raw material for natural selection that will favour the best adaptation to the environment, that of most benefit to the organisms themselves.

Left, the short-faced tumbler pigeon: 'In their extremely short, sharp, and conical beaks, with the skin over the nostrils but little developed, they almost depart from the type of the Columbidae.' (*Variation of Animals and Plants*, 1868)

Below, the pouter pigeon: 'The beak in one bird which I possessed was almost completely buried when the oesophagus was fully expanded. The males, especially when excited, pout more than the females, and they glory in exercising this power.' (Ibid.)

Although these conclusions were not new, being for the most part a corroboration of the basic theoretical lines he had set out in 1859, Darwin did introduce one new feature. This was a hypothesis on the internal workings of the hereditary mechanism which he developed in the penultimate chapter of the book, under the title 'Provisional Hypothesis of Pangenesis'. Seeking to solve the mystery of the hereditary transmission of characteristics, Darwin postulated that each part of the body contained a multitude of invisible particles, which he called 'gemmules'; these represented the part of the body in which they were found, and converged by affinity in the organs of reproduction to form the sexual elements, before uniting, also by affinity, with corresponding 'gemmules' from the second parent.

Although this unusual hypothesis was based within the framework of blended inheritance, and allows for the possibility of the transmission of 'acquired' traits, it also bears a

remarkable resemblance to the theories of generation developed in the previous century by Maupertuis and Buffon, in the wake of exported versions of the Newtonian model of the attraction of bodies and its great metaphors of the chemical affinity of substances. It also includes plants as well as animals.

Darwin and botany

Darwin produced a considerable amount of botanical work. In addition to articles, and whole chapters in *Origin of Species* (1859) and *The Variation of Animals and Plants* (1868), he devoted six major monographs to plants. He was not a classical botanist, paying little attention to plant systematics, for example, but in the course of his research he became an excellent plant chemist and physiologist, and was passionately interested in observation and experimentation.

Plants are subject to rapid variations that are easily propagated in controlled spaces, and so, as is well known, are a choice field for the observation of phenomena of change, and hence for 'transformist' interpretations. Darwin's theoretical idea was simple: it was to draw an analogy between the plant and animal kingdoms, as permanent evidence of the great kinship that unites all organisms within the complex and branching history of the living world. Like animals, plants have been partially domesticated, and have revealed themselves to be sensitive to changes in conditions, especially climatic ones; like animals – indeed, more so than animals – they vary, and their

The Venus flytrap (above) is an insectivorous plant, armed with formidable, jaw-like leaf-traps at the base of its stem, which can snap shut on captive insects in less than a second. Such examples of predatory behaviour seem to weaken the demarcation line between the plant and animal kingdoms. Below, sample boxes used by Darwin.

The Venus flytrap (left) comes from the marshlands of Carolina. It closes the lobes of its leaves in reaction to the touch of an insect, and the grip of the interlaced cilia along the lobe edges grows tighter the harder its prey struggles. This legendary plant is both disturbing and fascinating because it seems to possess an animal characteristic: motor coordination that produces an instaneous response. Below, an orchid.

variations can be selected; like animals, their tissues can be stimulated and contracted; like animals, they feed themselves, and several of them are even predatory (the insectivorous plants); like animals, they ingest, digest, excrete, sleep; like animals, they possess sexuality and can reproduce, either through self-pollination or through cross-fertilization, and pass on characteristics through inheritance. Like animals, but more easily, they can hybridize; like them, they can grow and move; they migrate (by means of the wind, or birds, sea currents, or, more recently, humans) and acclimatize; through unconscious or methodical selection, they can be made useful or pleasing to man; they can also beget deviant or monstrous individuals.

In the course of his research, Darwin studied and sketched the movements of plant tendrils, stems and leaves, investigated the mechanisms of pollination, demonstrated and explained the amazing co-adaptation of pollinizing insects and flowers, and noted the prevalence of cross-fertilization (between different individual plants) as opposed to

Two remarkable examples of mimicry for the purpose of camouflage, which have developed through natural selection. Opposite, an example of offensive mimicry: this bumblebee, *Bombus pratorum*, may seem to be gathering pollen, but has in fact been captured by *Thomisus onustus*, Europe's largest crab spider, whose coloured markings perfectly imitate the mauve shades of the scabious flower. Left, a striking example of protective mimicry: this African leaf grasshopper, *Brycoptera lobata*, exactly resembles an old, yellowing leaf, complete with scalloped edges that look as if they have been nibbled by some animal. Note the almost scarlike mark that runs along the edge of the 'eaten' area. The insect blends in with its habitat, becoming almost invisible to predators.

self-pollination (between male and female elements of the same individual). He concluded that there was an evolutionary advantage linked to sexual reproduction, cross-breeding and the diversity that it produces: this advantage is increased adaptability, leading to an increased ability to survive. With his son Francis in 1880, he performed a series of experiments on the partial lighting of plants and identified the phenomenon of action at a distance (transmission of a photosensitive stimulus) in individual plants.

Responding to criticism

From 1861 to 1872 (from the third to the definitive sixth edition) Darwin fine-tuned, in chapters VI and IX of *Origin of Species* and in chapters VI, VII and X

The coelacanth (*Latimeria chalumnae*), a bony fish of the order Actinistia, was long believed to be extinct, until a living specimen was captured in 1938 off the mouth of the Chalumna river, at Africa's southernmost point. Certain features of the Actinistia, such as the distribution of the pairs of fins which are articulated on a single base, resemble those of terrestrial tetrapods or vertebrates. However, it is the Dipneusti or

of the final edition, his responses to the criticisms that his theory had attracted. The first was the classic objection to Darwinian gradualism: if all species have evolved, one from another, by imperceptible degrees, then why are species so marked, and why do we almost never find among them a multitude of transitional forms? The answer is as follows: the forms favoured by selection eliminate and replace neighbouring forms that are related or in competition with them. Forms that display only slight variations

lunged fishes, another interest of Darwin's, which are today considered the closest relatives of the tetrapods.

have very little chance of reaching a relative permanence. In addition, varieties that appear in two different regions of one territory may see arise between them, in a thin border zone, an intermediate variety that is fewer in number, and which will consequently be less suited to producing selectable variations, and thus less fit to survive. In terms of fossils, the absence of intermediate forms can be explained by a formerly different distribution of land above sea-level (less continuous land, more islands, hence more distinct speciations), and also, doubtless, it is due to a scarcity of data in our geological archive. Geological collections can never be perfect, bearing in mind the impossibility of reconstructing interactions between ancient species, the extraordinary geological timescale involved, the alternation between periods of uplift and periods of subsidence, migrations, the shifting nature of delimitations between species and varieties, and especially the fact that we are mistaken in wishing to find forms that are directly intermediate between species, when what we should be seeking are forms intermediate between each species and a common but distinct ancestor that is unknown to us.

Another objection concerned the age of the Earth. William Thomson, the future Lord Kelvin, refused to accept that the Earth was old enough to allow the

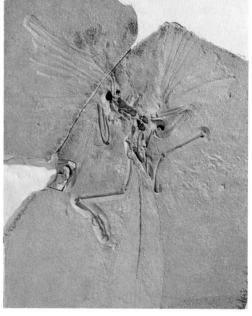

Below, archaeopteryx fossil. *Archaeopteryx lithographica* (von Meyer, 1861) was classed with the birds by Linnaeus (1758) and in the sub-class Sauria by Haeckel (1866).

To evolutionists, it is a precious relic of the path from reptiles to birds. The size of a pigeon, with feathers, teeth and claws, it has some skeletal features in common with the dinosaurs. Darwin discussed its discovery in *Origin* (chapter IX), and its value as evidence of transition in *Descent of Man* (chapter VI).

gradual unfolding of evolutionary change that Darwin alleged. He made this claim with the help of arguments taken from thermodynamics; the earth's insufficient degree of cooling today led him to conclude that the planet was between 20 and 200 million years old). Darwin merely referred Thomson to current ignorance about the globe's internal processes, and it was not until the 20th century, with the work of Pierre Curie and Ernest Rutherford, that this objection was finally swept aside: it is the loss of radioactivity in elements of the earth that explains the 'delay' in the cooling.

Sir William Thomson (1824–1907; below) was a collaborator of Joule, with whom he studied the cooling of gases, and a major theoretician on thermodynamics. Unlike Lyell and Darwin, he believed that thermodynamics could prove that the Earth was not old enough for evolutionary change to be possible.

Thomson's objection was also taken up by the Scottish engineer Fleeming Jenkin, who opposed Darwin with the idea that selection and gradual transformation could, within the framework of a theory of blending inheritance (producing characteristics intermediate between those of the two parents), override the dilution of the individual variations in cross-breeding. Darwin replied by pointing out the possibility that several lifeforms, and not necessarily just one, could be affected by the same variation. Jenkin's objection was in its turn swept aside in the next century by Mendelian genetics and the law of segregation of characteristics.

Finally, St George Jackson Mivart, a lawyer and zoologist, a convert to Catholicism and a friend of Owen, wrote a series of criticisms in 1871. In Darwin's eyes, the most serious of these was the following: natural selection distinguishes and promotes structurally useful variations, yet it can have no effect on variations of this type when nascent since, in Darwin's gradualist representation, at this early stage they cannot yet have attained any level of usefulness.

Darwin responded with what his follower Anton Dohrn was later to call the principle of succession

William Buckland (1784–1856) was a geologist and cleric, and twice president (1824, 1840) of the Geological Society, of which the young Darwin was secretary between 1838 and 1841. He was also Dean of Westminster from 1845, and curator of the British Museum from 1847. A strict natural theologian, he sought

of functions. Every organ is endowed with a principal function and a secondary function. Under the pressure of conditions, the ratio of functional importance can gradually be inverted, and the organ will undergo a process of transformation; each stage in the process will increase the usefulness of the organ without disturbing or destroying the functionality of the whole.

Mivart's second objection concerned what he claimed was the impossibility of constructing such a marvellously complex structure as a vertebrate's eye through the accumulation of selected variations, Darwin replied that each stage in the organ's growing complexity was selected precisely because of the growing degree of usefulness which it provided to creatures as they gradually became better adapted to their environments.

to reconcile the Earth sciences with the Biblical accounts of the Creation and the Flood. In 1840 Darwin read half of his book *Geology and Mineralogy considered with Reference to Natural Theology* (London, 1836), and re-read it in 1848. This Biblical geology was later swept away by the new geology, which was represented in England by Lyell's uniformitarianism. Above, an engraving of a cross-section of the Earth, taken from a work by Buckland.

'Important as the struggle for existence has been and even still is, yet as far as the highest part of man's nature is concerned there are other agencies more important. For the moral qualities are advanced, either directly or indirectly, much more through the effects of habit, the reasoning powers, instruction, religion, etc., than through Natural Selection; though to this latter agency may be safely attributed the social instincts, which afforded the basis for the development of the moral sense.'

The Descent of Man, 1871

CHAPTER 5

NATURE AND CIVILIZATION

Darwin discusses anthropology in a major book published in 1871, *The Descent of Man, and Selection in relation to Sex*. Opposite, Darwin in 1881, a year before his death. Below, a chimpanzee.

Sexual selection and altruism

More than eleven years after the first edition of *Origin of Species*, which illustrated the evolution of life and proposed an explanation for it through the effects of natural selection on the animal and plant kingdoms, Darwin finally tackled the human world in his great book *The Descent of Man, and Selection in relation to Sex* (1871).

Its fundamental aim was to demonstrate man's genealogical connection with the animal kingdom, and more precisely to highlight his descent from an ancestor related to the Catarrhine monkeys of the Old World – and hence, against the last bastions of dogma, to deliberately extend 'transformism' to the human species. At the same time, it attempted to explain the evolution of the human species, and prove that natural selection had also modelled our biological history. Finally, it aimed to show that human civilization, which protects the weak rather than exterminating them, is diametrically opposed to the elimination stage of the selective process which, through methods which are also variable and evolving, had been in operation up to that point.

Natural selection not only works to select advantageous organic variations, but also instincts. Throughout the animal kingdom, it has selected social instincts, whose evolutionary fusion with rational intelligence culminated in humankind; within us these instincts are accompanied by the indefinite extension of the influence of morality and altruism. Favoured by selection, social

The closeness of man and ape gave rise to countless caricatures playing on the humanization of the ape and at the same time, on the animalization of Darwin (above). The similarities between humans and apes are not restricted to anatomical structure, but encompass instincts, social organization and behaviour, including imitation, rational faculties, the ability to learn and mutual aid, all powerful evolutionary factors. Left, an ape skull that belonged to Darwin.

instincts, together with rationality, have changed human evolutionary history by favouring anti-selective behaviour: moral education, care for the sick and disabled, measures to care and compensate for physical handicaps and learning disabilities, the institutionalization of charity and social intervention on behalf of the underprivileged.

This movement of gradual reversal has been labelled 'the reversive effect of evolution' (Tort 1983): through social instincts, natural selection chooses civilization, which runs counter to natural selection. The advantage obtained is no longer a biological one: it has become a social one. This apparent discontinuity without any actual discontinuity is the theoretical key to Darwin's materialist theory of continuity in his consideration of the relationship between nature and civilization.

Above, mothers take their families to visit a 'distant relative'. Contrary to popular belief, Darwin was neither misogynist nor sexist. Although he explained the inferior status of women in most societies in evolutionary terms, he conversely established that, as possessors of the most fundamental of social instincts (maternal love, basis of the protection of the weak), they embody the ethical future of humanity and the concept of 'civilization'.

The first vehicle for altruism is the instinct to copulate and procreate: during the courtship dances that precede mating in many animal species, the male is often burdened with a handicap of secondary sexual characteristics (horns, weapons, feathers or other ornamental features) which he uses either in combat or in the courtship of females; in other words, to lure a potential companion and mate in his direction. However, these ornaments are potentially deadly. Covered in his splendid but cumbersome finery, the bird of paradise might appear irresistible to females, but is almost incapable of flying and is therefore in great danger from predators. Females, meanwhile, will lavish care on their young, and may even put themselves in danger in order to protect their progeny. Social instincts, therefore, are seen to have an evolutionary history which even includes the concept of self-sacrifice; this reaches its culmination in human morality. Darwin succeeded in producing a genealogy of morals without any reference to a supernatural authority.

If the earliest seed of altruism is linked to fertilization, then the embryonic form of social instinct (and therefore, eventually, of morals) is found in the care of offspring, which is usually the privileged function of the females of a species. As far as human beings are concerned, Darwin put a great emphasis on the moral superiority of women, who have more direct access than men to the emotions of tenderness and compassion.

The magnificent display plumage of the male bird of paradise (*Paradisea apoda*; left) is both a force in the process of sexual selection, and a potentially fatal weakness. The link between the offering of love and the risk of death seems to foreshadow aspects of the human mythology of love.

•Cats, when terrified, stand a full height, and arch their backs in a well-known and ridiculous fashion. They spit, hiss, or growl. The hair over the whole body, and especially on the tail, becomes erect.•
The Expression of the Emotions in Man and Animals
[Below, engraving of a frightened cat by Wood.]

Expressions and emotions

In 1872, Darwin published *The Expression of the Emotions in Man and Animals*. Responding to the creationist and finalist book by Charles Bell (1774–1842), *The Anatomy and Philosophy of Expression* (1806), which claimed that the facial muscles were specially created for the purpose of expressions, Darwin demonstrated the similarities of reactive behaviour and expressions in animals and humans, and distinguished three principles that control their operation:

1. The principle of the association of useful habits. For example: because of an involuntary association, a man may recoil from the attack of a snake from which he is protected by a window. The voluntary inhibition of this act is often accompanied by movements of contraction, which are also expressive;

2. The principle of antithesis. A sudden change from a potentially 'useful' attitude to an inverse state triggers involuntary expressive movements that are not useful, but are the complete opposite of those that were previously displayed. For example, a dog growling at a stranger may suddenly switch to happiness and affectionate tail-wagging when it recognizes its own master;

3. The principle of the direct action of the nervous system. An excess of nervous energy caused by violent excitement is expressed or, alternatively, suddenly interrupted. Example: nervous trembling and faster heartbeat during periods of acute emotion.

French doctor Guillaume Duchenne (1806–75) carried out a major study of the expression of the emotions, and Darwin used the photographs from Duchenne's book

Mécanisme de la physionomie humaine (1862) showing different contractions of the facial muscles when stimulated by electrical impulses (above). However, like Charles Bell, Duchenne viewed this as a universal physiognomic language, given to man by God, a notion that Darwin considered naive.

This fine book – Darwin's 'hobby-horse', as he called it in a letter to Wallace – discreetly lays the foundations of comparative animal psychology and ethology in the field of modern evolutionary theory. Some ten years before his death, despite suffering from almost permanent ill health, Darwin had already produced a considerable body of work. The best known of his books (*Origin*, *Variation*, *Descent*, *Expression*) resemble huge, interconnected chapters of his planned but uncompleted 'Big Book on Species', of which large sections remain among his notes and manuscripts.

Looking at the earth

In 1876 Darwin wrote his *Autobiography* for his family and friends. In 1879 he added a biographical 'Preliminary Notice' to the English edition of Ernst Krause's book about his grandfather, Erasmus Darwin. The publication of the great botanical monographs would end in 1880. The following year, Darwin took a strong line on vivisection, which he defended as experimentation that was vital to the progress of medical science, although he called for the greatest humanity towards animals. His last book, *The Formation of Vegetable Mould, through the Action of Worms, with Observations on their Habits* (1881), highlighted the important role these creatures have played throughout geological history: the recycling, refinement, aeration and breaking up of soil, the decomposition of leaves and formation of humus, the disintegration of rock particles, the production of a large amount of friable evacuations carried off by water and wind, and the burial of bodies left on the surface. He even noted the evidence for intelligent behaviour among worms, particularly their procedures for sealing galleries. Once again, at the heart of this book lies Darwin's fascination with geology. As the basis of his method,

This Punch caricature shows Darwin pondering the actions of worms, one of which is springing from his brain like a giant question mark. With hindsight it might seem like a meditation on his imminent death. In his 1881 book, Darwin examined the fate of Silchester Basilica, whose burnt ruins were criss-crossed by worms. The worms brought topsoil to the surface and covered the ruins, showing that through destruction, biological cycles continue. A year later, Darwin died and was buried in Westminster Abbey.

the direct observation of micro-processes in action allowed him to understand the macro-processes which have changed the world over time through a slow process of accumulation.

In 1882, Darwin pursued his research into the action of chemicals on plant tissue, and studied the geographical dispersal of freshwater bivalves, and animal behaviour. He wrote a preface for the English edition of *Studies in the Theory of Descent* by August Weismann, founder of neo-Darwinism, who theorized the independence of germ (or sex) cells and somatic (or body) cells, thereby pioneering genetics and refuting the idea of Lamarckian inheritance of acquired characteristics. Darwin also provided a preface for *The Fertilisation of Flowers* by Hermann Müller, brother of Fritz Müller, a German naturalist who had emigrated to Brazil, and author of a famous defence of Darwinian theory, *Für Darwin* (1864).

But Darwin grew weak and, on 19 April, at half past three in the afternoon, he died peacefully at Down House. His family, close friends and a host

Left, the Worm Stone, placed on the lawn of Down House to measure the movements of the soil caused by earthworms, and especially the speed of burial. This gadget was reconstructed in 1929 by the Cambridge Instrument Company which was managed by Horace, one of Darwin's sons. In the conclusion of his 1881 book, Darwin wrote: 'Hensen placed two worms in a vessel 18 inches in diameter, which was filled with sand, on which fallen leaves were strewed; and these were soon dragged into their burrows to a depth of 3 inches. After about 6 weeks an almost uniform layer of sand, a centimetre (0.4 inch) in thickness, was converted into humus by having passed through the alimentary canals of these two worms.'

This experiment demonstrates the production of vegetable mould through the digestive activity and evacuations of worms: a classic example of the biological recycling of the earth.

of names from the worlds of science, politics and even religion accompanied his body to Westminster Abbey, where, contrary to his wishes, he was buried on 26 April, in the presence of both admiring supporters and respectful adversaries who wished to pay homage to a great gentleman of science. He never doubted the truth of his theory, but he did doubt his own ability to convince. Nonetheless, his struggle succeeded in establishing 'transformism' as a solid theory during his lifetime, although it was another century before Darwinism, rid of the ideological dross that had distorted its image, would be understood in all its powerful and complex originality.

Although his struggle against religious ideology and creation mythologies never seemed like a direct polemical confrontation, something he had always avoided as potentially doing more harm than good, it was nevertheless constant, and ran parallel to his own changing beliefs and to what his work had taught him about religion itself as an evolving phenomenon.

Darwin in the drawing room of Down House, towards the end of his life, listening to his wife play the piano. 'Only picture to yourself a nice soft wife on a sofa with good fire, and books and music perhaps,' he had scribbled on some old paper during the deliberations which preceded his marriage. At Cambridge, Charles was already paying choristers to sing in his apartments, and his taste for music, an art for which he had no talent of his own, remained a mystery to him. Emma played for him, and also read at his bedside when his eyes were too tired.

NATURE AND CIVILIZATION 101

Darwin and religion

When he departed on the *Beagle*, Darwin, although
born of a double lineage of agnostic thinkers, had
never thought to question his religious education or
Paley's natural theology, which he had studied at
Cambridge at the time when, although lacking in
conviction, he had planned to become a parson. Since
everything in nature seemed to him to be orientated
towards an end (he was a finalist), the works of nature
appeared to obey an intelligent plan conceived by a
supreme creative intelligence (he was a creationist),
and were all simply proofs of providence (he was a
providentialist). But he wanted to understand, and
refused to sacrifice his intelligence to a faith built
on his failure to do so.

On his return, Darwin had changed. He was
already convinced by 1837 that species were
transformed over the course of the Earth's history,
and therefore that the first book of the Bible, Genesis,

Darwin's funeral at
Westminster Abbey
on 26 April 1882. A
suggestion by Galton,
supported by members
of the X Club, a group
of Darwin's friends,
and by the academic
and parliamentary
authorities, was able
to persuade his family,
as well as the religious
and political figures in
charge, to consent to the
ceremony. One of the
greatest participants in
freeing science from
God duly received the
stiff homage of the
Church of England and
conservative circles.

which tells of the creation of the world and all living species by God in six days, must be false or deceitful. From then on, he gradually began to break away from religion. This was not a painless process, but it was an irrevocable one. He was tempted by deism, a personal belief free of dogma, based on the idea that the harmony of the world could not be the result of chance alone. But the theory of natural selection brought him the solution to this problem by providing an internal mechanism that was capable of producing and sustaining the order and balance of the world. 'Thus', he wrote in his *Autobiography* of 1876, 'disbelief crept over me at a very slow rate, but was at last complete.'

He did not believe in miracles or in the barbaric idea of a tyrannical and cruel God, and thus rejected certain passages of the Old Testament and Christian morality. He saw these beliefs as the last traces of a primitive state of being; this prompted him to consider the history of religion itself as an evolving process. He also condemned the idea of eternal punishment for non-believers. He imagined the existence of nations without religion, and put the value of religion into perspective by showing that the peoples of world all have different beliefs and traditions. He contrasted the infinite suffering of living beings with the image of an omnipotent and benevolent God. The immorality of unjust suffering is more in keeping with natural selection, which is morally 'blind', than with Providence. Not wishing to shock the still dominant convictions of his contemporaries and compromise his work as a result, he formally reintroduced a mention of the Creator at the end of the second edition of *Origin of Species*, in January 1860.

On the outside, he maintained an attitude of agnosticism; knowing that there are many things that cannot be

known, he never declared that God did not exist, and refused to get involved in theological or metaphysical disputes, only confronting the Church when its representatives opposed the advancement of scientific ideas. But inside he was a materialist and a non-believer. As an observer of animals, he even detected some 'animist' behaviour in them (the attribution of a force, 'soul' or 'spirit' to inanimate objects, beyond their material appearance), a primordial manifestation of religious belief. Darwin believed that religion, like morality, is simply another product of evolution; its stages can be retraced and its precepts rediscovered by paths other than that of divine revelation.

In October 1882, six months after his death, his wife Emma, whose religious beliefs he had always respected, had removed from his *Autobiography* the passage in which Darwin declares as definitively reprehensible the idea of eternal punishment for unbelievers. This passage was restored by Nora Barlow, his grand-daughter, in 1958.

The Biblical story of the Creation (Genesis 1: 24-25) tells that, during the sixth and final day of his work, God created Man in his own image, and each domestic animal 'after his kind.' Then God placed him in the Garden of Eden to dress it and to keep it (2: 15). Above, Jan Brueghel the Elder's *Earthly Paradise*.

The former traveller settled at Down House (left, the verandah and park) at the age of 33, and only left it for occasional visits and several cures. Darwin received many eminent guests here: Lubbock was a neighbour, and regular visitors included Edward Forbes, the mollusc specialist, Thomas Bell, Hooker, Huxley, Lyell, Romanes (after 1874), Wallace and Waterhouse. Henslow came in 1854; FitzRoy in 1857; physiologist William Benjamin Carpenter in 1861; Swiss anatomist and embryologist Rudolf Albert von Kölliker in 1862; Haeckel in 1866; the young Russian palaeontologist V. O. Kovalevsky in 1867 and 1870; Bates, Blyth and Gray in 1868; Alexander Agassiz (son of Louis) in 1869; Chauncey Wright in 1872; Anton Dohrn in 1873; Gladstone, the former (and future) Liberal prime minister in 1876; Edmond Barbier, the French translator of part of his work, and Augustin Pyrame de Candolle in 1880; and Edward Aveling, companion of Eleanor Marx and a free-thinker who reconciled Darwinism and Marxism, in 1881, accompanied by the materialist theoretician Ludwig Büchner.

A living theory of civilization

Modern biology owes its intelligibility to Darwin. Since every living thing is a product of evolution, the study of evolution must be the first step towards an overall understanding of the diversity and unity of biological phenomena. With Darwin's responses to criticism from 1860 onward, with Weismann and neo-Darwinism, with the development from 1900 onwards of Mendelian genetics and then, in the 1930s and 1940s, with the advent of the synthetic theory of evolution, and more recent criticisms of the latter, the history of Darwin's theory of the evolution by means of natural selection has been punctuated by eclipses and temporary crises. But from each of these, it has emerged improved and enriched, as is the rule for every living scientific theory.

Palaeontology, molecular biology, population genetics and modern systematics are still today confirming some findings and bringing new perspectives to Darwinism. Despite ongoing attacks by fundamentalist creationists (Kansas in 1999) and providentialist reinterpretations of evolution by mystic groups or churches themselves, it remains the great theoretical framework of the contemporary scientific study of the facts of evolution.

In more general terms, the theory of civilization which was developed in *The Descent of Man*, rediscovering the common origin and the joint evolution of social instinct, communal rationality and moral feelings, renews the link between nature

Along with Austrian agronomist Erich Tschermak and German botanist Carl Correns, the Dutch botanist and cytologist Hugo De Vries (1848–1935; below) was one of the 'rediscoverers' of Mendel's laws in 1900 and therefore one of the founders of modern genetics, which is opposed to the idea of inheritance of acquired modifications. It was the sudden mutation of a flowering plant, *Oenothera lamarckiana*, which in 1886 pointed him towards the discovery of the laws of heredity and a theory of evolution by leaps (saltationism). Opposite, Darwin with his orchids, in the hothouse at Down House.

and culture, between eliminatory selection and the institutionalization of assistance for the weak, without breaking the necessarily uninterrupted thread of evolutionary continuity. The key to Darwin's anthropology lies in the formula that sums up the so-called 'reversive effect of evolution': natural selection selects civilization, which runs counter to natural selection. This anthropology is persistently and sometimes vehemently opposed to the ideologies that have tended to give credence to the idea of a natural basis for the conduct of oppressive

domination of the strong over the weak, of the fit over the less fit, of the able-bodied over the suffering, the rich over the poor, one race over another, the 'civilized' over the 'savage', the 'superior' over the 'inferior', because it starts out by refusing to accept that the notions of inferior and superior are hierarchically innate.

In fact, if 'superiority' and 'inferiority' were intrinsic and permanent attributes of individuals or species, evolution would not be possible. The random variability of organisms and the fluctuating nature of the environment, either separately or together, mean that it is impossible for any vital advantage to be permanent, and equally impossible for it to be innately superior in any way. A simple chemical change in the environment can mean that a species which has until that point been in a minority, however negligible, can become dominant if it

The Möbius strip (left) illustrates the concept of reversive evolution. If a strip with two surfaces is closed after half a twist, it only has a single surface. If the two originally opposite surfaces are called 'nature' and 'civilization', it becomes possible to pass from one to the other without a jump or a break (there could not be any in a genealogy). The Darwinian theory of continuity is not simple but reversive. The movement from nature to culture does not produce a break but merely the effect of a break, because the movement 'to the other side' occurs gradually. This figure can also be used to illustrate that the sciences of civilization (the social sciences) need not contradict humanity's biological 'nature', but are nonetheless distinct in subject and method from human biology.

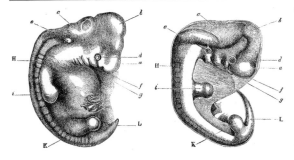

These diagrams from the first chapter of *The Descent of Man* show the remarkable resemblance between human (left) and dog (right) embryos at an early stage of development, indicating an ancestral kinship. Below, the cover of a modern French edition of *The Descent of Man* uses an image of mating butterflies to illustrate the theme of sexual dimorphism and selection, those 'motors' of evolution which cause apparently altruistic behaviour and sometimes seem to contradict the rules of natural selection.

happens to possess traits that will enable it to survive and prosper within the changed environment.

However, Darwin did more than warn us about the relative and provisional nature of any hierarchy. From the simple basic idea of the selection of social instincts, he constructed a theory of the immanent generation of morality, a materialist genealogy of morals. This theory, based solely on the dynamic of selection, highlights the fact that natural selection includes the possibility of self-application, and clinches the overall understanding of the relationship between nature and civilization from the point of view of the theory of descent through modifications: natural selection is not only the law or motor of evolution, it is also in itself an evolving reality, and therefore subject to its own law, that of evolutionary divergence.

The model of natural selection demonstrates, through the branching genealogical tree of the living world, that the evolutionary success of a new variation is usually accompanied by the dying out of the form from which it stemmed. Following the same model of development, within natural selection itself a major substitution gradually occurs: the progressive replacement of an older form of selection that favours

sous la direction de Patrick Tort
ŒUVRES DE
CHARLES DARWIN
Traduit, sous dimoix par
Michel Prum
précédé de
PATRICK TORT
L'anthropologie inattendue de Charles Darwin

LA
FILIATION
DE
L'HOMME
ET LA SÉLECTION LIÉE AU SEXE

the elimination or exclusion of the weak or the less fit, by a variant and better adapted mode of selection that favours altruistic and interdependent modes of behaviour which the group, as it slowly reaches an increasingly high level of social complexity, begins to use to aid its own weaker members. This is not a conflict with nature, because it is known that forms of behaviours such as mutual aid, cooperation and even charity are quite usual in some areas of the animal kingdom, wherever they have come to represent an evolutionary advantage. Nor is this – as was claimed by the theory of continuity of sociobiology of the 1970s – merely a tactic to ensure the increased survival of the genes, a process in which individuals are unconscious vehicles. Through the strength of their principles, the ethics of charity and aid seem to absolutely transcend any individual, populational or ethnic interests, and continue to function without any regard to probable genetic relationships or distances.

A century before the great contemporary socio-anthropological and biological debates, Darwin tackled the question of the relationship between nature and civilization, and opposed both the believers in a discontinuity at a cultural level and the supporters of biological continuity at all costs. The issue is sometimes raised even today, although Darwin himself rendered it null and void. His solution was still based entirely on the biological theory of continuity linked to the genealogical perspective of 'transformism'. Nevertheless, it takes into account the fact that

In the animal kingdom, the task of caring for and protecting offspring is mainly carried out by females, as can be seen among members of such different species as centipedes (opposite left) and apes (left). Darwin used the terms 'maternal instinct' and 'maternal love' interchangeably, in order to highlight the most basic form of social altruism and moral feelings inherent in this early demonstration of protection of the weak.

The study at Down House (overleaf). Above the mantelpiece hang three framed portraits: Charles Lyell, given to Darwin by Lady Lyell in 1847; Joseph Dalton Hooker, a gift from the photographer Julia Cameron; and Josiah Wedgwood II or 'Uncle Jos' (the only one visible here), which was restored to its place in Down House as a gift from Francis Darwin in 1927.

evolution has the ability, inherent in its basic dynamic of variation, selection and divergence, to encompass within itself not a schism, but the appearance of a schism. This makes it possible in future for humankind to disrupt the ancient law of selection, that of the triumph of the fittest, of elimination and inflicted death, and to establish in its place, in compliance with the moral standards passed on by education and institutions, a mode of anti-selective behaviour that forms the very heart of civilization. Here, in the truest sense and full of ethical implications, lies the real dialectic of nature.

DOCUMENTS

Erasmus Darwin's *Zoonomia*

The thread of many of Darwin's ideas can be seen in the work of his grandfather Erasmus, a physician, poet and philosopher. Zoonomia; or the Laws of Organic Life *was one of the first formal theories on evolution and described the common source of all life as a single 'living filament'.*

...When we revolve in our minds, first, the great changes, which we see naturally produced in animals after their nativity, as in the production of the butterfly with painted wings from the crawling caterpillar; or of the respiring frog from the subnatant tadpole; from the feminine boy to the bearded man, and from the infant girl to the lactescent woman....

Secondly, when we think over the great changes introduced into various animals by artificial or accidental cultivation, as in horses, which we have exercised for the different purposes of strength or swiftness, in carrying burthens or in running races; or in dogs, which have been cultivated for strength and courage, as the bulldog; or for acuteness of his sense of smell, as the hound and spaniel; or for the swiftness of his foot, as the greyhound; or for his swimming in the water, or for drawing snow sledges, as the rough-haired dogs of the north; or lastly, as a play-dog for children, as the lap-dog; with the changes of the forms of the cattle, which have been domesticated from the greatest antiquity, as camels, and sheep; which have undergone so total a transformation, that we are now ignorant from what species of wild animals they had their origin....

Thirdly, when we enumerate the great changes produced in the species of animals before their nativity; these are such as resemble the form or colour of their parents, which have been altered by the cultivation or accidents above related, and are thus continued to their posterity. Or they are changes produced by the mixture of species as in mules; or changes produced probably by the exuberance of nourishment supplied to the fetus, as in monstrous births with additional limbs; many of these enormities of shape are propagated, and continued as a variety at least, if not as a new species of animal....

Fourthly, when we revolve in our minds the great similarity of structure which obtains in all the warm blooded animals, as well quadrupeds, birds, and amphibious animals, as in mankind; from the mouse and bat to the elephant and whale; one is led to conclude, that they have alike been produced from a similar living filament. In some this filament in its advance to maturity has acquired hands and fingers, with a fine sense of touch, as in mankind. In others it has acquired claws or talons, as in tygers and eagles. In others, toes with an intervening web, or membrane, as in seals and geese. In others it has acquired cloven hoofs, as in cows and swine; and whole hoofs in others, as in the horse....

Fifthly, from their first rudiment, or primordium, to the termination of their lives, all animals undergo perpetual transformations; which are in part produced by their own exertions in consequence of their desires and aversions, of their pleasures and their pains, or of irritations, or of associations; and many of these acquired forms or propensities are transmitted to their posterity....

As air and water are supplied to animals in sufficient profusion, the three great objects of desire, which have changed the forms of many animals by their exertions to gratify them, are those of lust, hunger, and security. A great want of one part of the animal world has consisted in the desire of the exclusive possession of the females; and these have acquired weapons to combat each other for this purpose, as the very thick, shield-like, horny skin on the shoulder of the boar is a defence only against animals of his own species, who strike obliquely upwards, nor are his tushes for other purposes, except to defend himself as he is not naturally a carnivorous animal....

Another great want consists in the means of procuring food, which has diversified the forms of all species of animals. Thus the nose of the swine has become hard for the purpose of turning up the soil in search of insects and of roots. The trunk of the elephant is an elongation of the nose for the purpose of pulling down the branches of trees for his food, and for taking up water without bending his knees. Beasts of prey have acquired strong jaws or talons. Cattle have acquired a rough tongue and a rough palate to pull off the blades of grass, as cows and sheep. Some birds have acquired harder beaks to crack nuts, as the parrot. Others have acquired beaks adapted to break the harder feeds, as sparrows. Others for the softer seeds of flowers, or the buds of trees, as the finches.... All which seem to have been gradually produced during many generations by the perpetual endeavour of the creatures to supply the want of food, and to have been delivered to their posterity with constant improvement of them for the purposes required.

The third great want amongst animals is that of security, which seems much to have diversified the forms of their bodies and colour of them; these consist in the means of escaping other animals more powerful than themselves. Hence some animals have acquired wings instead of legs, as the smaller birds, for the purpose of escape. Others great length of fin, or of membrane, as the flying fish, and the bat. Other great swiftness of foot, as the hare. Others have acquired hard or armed shells, as the tortoise and the echinus marinus....

From thus meditating on the great similarity of the structure of the warm-blooded animals, and at the same time of the great changes they undergo both before and after their nativity; and by considering in how minute a portion of time many of the changes of animals above described have been produced; would it be too bold to imagine, that in the great length of time, since the earth began to exist, perhaps millions of ages before the commencement of the history of mankind, would it be too bold to imagine, that all warm-blooded animals have arisen from one living filament, which THE GREAT FIRST CAUSE endued with animality, with the power of acquiring new parts attended with new propensities, directed by irritations, sensations, volitions, and associations; and thus possessing the faculty of continuing to improve by its own inherent activity, and of delivering down those improvements by generation to its posterity, world without end?

Erasmus Darwin
Zoonomia, 1803

Evidence for evolution

Darwin's main concern in Origin of Species *was to demonstrate how living things are transformed through the mechanisms of variation and selection. Along with biogeography and palaeontology, he also used evidence from other scientific fields to illustrate his argument.*

A. MORPHOLOGICAL AND ANATOMICAL EVIDENCE

1. Unity of type, connection and homology

Darwin writes in 1844 (second outline of *Origin of Species*, chap. VIII) that within each large class of animals, across enormous differences in climate and time, and despite their very specific adaptations to environmental conditions, organisms possess a uniformity of structure that the greatest naturalists have all recognized.

Following Goethe's famous treatise on the metamorphosis of plants, Etienne Geoffroy Saint-Hilaire laid down the central concepts of connection, analogy (Owen's homology) and unity of plan in *Philosophie anatomique* in 1818. The basic concept is that the limbs of each major group can be morphologically reduced to a single organizational plan. This unity of plan goes beyond any consideration of the size, shape, and special function of the organs, which are due to special adaptations, and refers instead to common peripheral structures, as well as the number and position of their constituent parts. According to the 'principle of connections', a lizard's foot seems to be built from the same design

and the same parts as a horse's hoof, a bird's wing or even a whale's flipper. In anatomical terms these are 'homologous' structures; that is to say that they are linked by a 'true' affinity and not by a simple convergent resemblance due to shared form and function of an organ, such as the case of a whale's flipper and a fish's fin, a similarity which is merely an 'adaptive character'. It is obvious that these concepts and distinctions have serious implications for 'transformism', because they allow for the supposition that increasingly differentiated forms have been created through a process of divergence and adaptation from a smaller number of ancestral forms that were more uniform.

In the same text, Darwin explores the inspiration for his theory: 'What, for instance, is more wonderful than that the hand to clasp, the foot or hoof to walk, the bat's wing to fly, the porpoise's fin to swim, should all be built on the same plan? And that the bones in their position and number should be so similar that they can all be classed and called by the same names. Occasionally some of the bones are merely represented by an apparently useless, smooth style, or are soldered closely to other bones, but the unity of type is not by this destroyed and hardly rendered less clear. We see in this fact some deep bond of

Homologous bone structures in the forelimbs of the dugong and the bat.

union between the organic beings of the same great classes – to illustrate which is the object and foundation of the natural system. The perception of this bond, I may add, is the evident cause that naturalists make an ill-defined distinction between true and adaptive affinities.'

Creationist and 'fixist' naturalists recorded these similarities and modifications in their descriptions and classifications, but could account for them only by reiterating the overall suitability of the organism for the environment in which it lives, and by metaphysical reference to final causes. The theory of descent through modification, on the other hand, explains these features as the result of the natural selection of advantageous variations to fit different environments. For example, a foot can be selected so that its bones gradually grow elongated and are connected by means of increasingly wide membranes, until it becomes a paddle for swimming. Darwin's active scientific interpretation

for this phenomenon was based on an understanding of the processes at work, rather than being a recitation of religious doctrine that does not adequately explain anything.

'By unity of type is meant that fundamental agreement in structure which we see in organic beings of the same class, and which is quite independent of their habits of life,' he wrote in *Origin of Species*. 'In my theory, unity of type is explained by unity of descent.' (chap. VI, summary). Another difficulty for teleologists was explaining the existence of what are known as vestigial or rudimentary structures and organs, since there seems to be an unfathomable contradiction between their presence, their 'finalization' for a given function, and their uselessness for that intended purpose.

2. Rudimentary organs

This is the name given to parts of an organism that seem to be regressed; they are not able to fulfil their apparent function and seem physiologically superfluous. They are also known as 'vestigial' organs, because since Darwin's time they have been recognized as residual vestiges of organs that long ago had a size, function and role that meant they were a normally integrated part of the whole organism. Classic examples include the toothbuds of whale embryos, the upper incisors of ruminant animals, the two lateral toes of the horse's foot or the appendix in humans: a small cylindrical organ, about 9 to 10 cm (3½ to 4 in.) long, located at the junction of the large and small intestines, which forms a closed sac and serves no obvious purpose.

In Darwin's view, a rudimentary organ was first defined by its apparent lack of purpose in the adult animal: 'In the mammalia, for instance, the males possess rudimentary mammae; in snakes one lobe of the lungs is rudimentary; in birds the "bastard-wing" may safely be considered as a rudimentary digit, and in some species the whole wing is so far rudimentary that it can not be used for flight. What can be more curious than the presence of teeth in foetal whales, which when grown up have not a tooth in their heads; or the teeth, which never cut through the gums, in the upper jaws of unborn calves?' (*Origin of Species*, chap. XIV, sixth and final edition).

Darwin's interest in rudimentary organs was closely linked to his theory of 'transformism'. An organ that is redundant – or at least does not fulfil its apparent function – cannot be explained by the present, and so must be interpreted with regard to the past. Its presence suggests that there was once an ancestral form of the organism in which the organ functioned normally. This induction implies that all

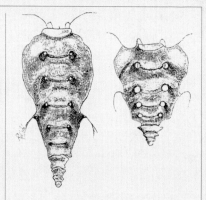

Rudimentary tail bones in the gorilla (left) and man. The coccyx serves to illustrate ancestral kinship and the process of regression.

organisms have been transformed over time, and provides valuable clues to their classification, which should be approached from a genealogical perspective:

'No one will say that rudimentary or atrophied organs are of high physiological or vital importance; yet, undoubtedly, organs in this condition are often of much value in classification. No one will dispute that the rudimentary teeth in the upper jaws of young ruminants, and certain rudimentary bones of the leg, are highly serviceable in exhibiting the close affinity between ruminants and pachyderms. Robert Brown has strongly insisted on the fact that the position of the rudimentary florets is of the highest importance in the classification of the grasses.' (*Origin of Species*, chap. XIV).

For example, the foetus of the land salamander does not live in water, but it possesses gills identical to those of aquatic salamander larvae, indicating that these two species share a single aquatic form as a common ancestor.

Rudimentary organs are also strong evidence against providentialist or finalist interpretations of nature, such as natural theology. A rudimentary organ seems to be 'designed' for a purpose, but this 'design' fails, thus contradicting the doctrine of the perfection of the divine plan. Darwin also responded to the Swiss naturalist Louis Agassiz who was obliged to recognize the uselessness of these organs but attributed them to God's concern for aesthetics:

'In works on natural history, rudimentary organs are generally said to have been created "for the sake of symmetry", or in order "to complete the scheme of nature". But this is not an explanation, merely a re-statement of the fact. Nor is it consistent with itself: thus the boa constrictor has rudimenta of hind limbs and of a pelvis, and if it be said that these bones have been retained "to complete the scheme of nature", why, as Professor Weismann asks, have they not been retained by other snakes, which do not possess even a vestige of these same bones? What would be thought of an astronomer who maintained that the satellites revolve in elliptic courses round their planets "for the sake of symmetry", because the planets thus revolve round the sun?' (Origin of Species, chap. XIV).

Natural selection selects adaptations that are useful to an organism, and so if an environmental change means that an organ is no longer useful, it will become rudimentary, since it is no longer serves a purpose. In some cases, a positive adaptation will be subjected to a process of regressive growth: 'An organ, useful under certain conditions, might become injurious under others, as with the wings of beetles living on small and exposed islands; and in this case Natural Selection will have aided in reducing the organ, until it was rendered harmless and rudimentary.' (Origin of Species, chap. XIV).

Darwin believed that rudimentary organs are valuable genealogical records. To illustrate this fact, he used one of Lyell's favourite linguistic analogies, one which also appears in 1916 in Ferdinand de Saussure's Cours de linguistique générale: 'Rudimentary organs may be compared with the letters in a word, still retained in the spelling, but become useless in the pronunciation, but which serve as a clue for its derivation.' (Origin of Species, chap. XIV).

Darwin wrote a very long letter to Lyell on 11 October 1859, in reply to two previous letters which had contained Lyell's responses to a proof version of Origin of Species. This particular passage is notable as it subordinates the major distinction between rudimentary and nascent organs to the principle of utility, a central concept in the theory of natural selection:

'On theory of Natural Selection. There is wide distinction between rudimentary organs and what you call germs of organs and what I call in my bigger book, "nascent" organs. An organ should not be called rudimentary unless it be useless, – as teeth which never cut through the gums – the papilla representing the pistil in male flowers – wing of Apteryx, or better, little wings under soldered elytra. These organs are now plainly useless, and a fortiori they would be useless in a less developed state. Natural Selection acts exclusively by preserving successive slight, useful modifications, hence natural selection cannot possibly make a useless or rudimentary organ. Such organs are solely due to inheritance (as explained in my discussion) and plainly bespeak an ancestor having the organ in a useful condition. They may be, and often have been worked in for other purposes; and then they are only rudimentary for the original function, which is sometimes plainly apparent. A nascent organ, though little developed, as it has to be developed, must be useful in every stage of development. As we cannot prophesy we cannot tell what organs are now nascent; and nascent organs will rarely have been handed down by certain members of a class from a remote period to present day, for beings with any important organ but little developed will generally have been supplanted by their descendants with the organ well developed. The mammary glands in Ornithorhynchus may perhaps be considered as nascent compared with the udders of cow. – ovigerous frena in certain cirripedes are nascent branchiae. – in Ameiva the swim-bladder is almost rudimentary for this purpose, and is nascent as a lung. The small wing of Penguin, used only as a fin might be nascent as a wing; not that I think so; for whole structure of bird is adapted for flight, and a penguin so closely resembles other birds that we may infer that its wings have probably been modified and reduced by natural selection in accordance with its sub-aquatic habits. Analogy thus often serves as guide in distinguishing whether an organ is rudimentary or nascent. I believe os coccyx gives attachment to certain muscles, but I cannot doubt that it is a rudimentary tail. The bastard-wing of birds is rudimentary digit; and I believe that if ever fossil birds are found very low down in series, they will be seen to have a double or bifurcated wing. Here is a bold prophecy! To admit prophetic

germs is tantamount to rejecting theory of Natural Selection.'

B. TAXONOMY AND CLASSIFICATION

1. Classification problems

When Darwin decided to become a naturalist, the classification of living things – plant and animal systematics – had been considered since antiquity to be one of the greatest tasks of natural history. The importance of taxonomy was due to the belief that within nature, all forms of life are objectively divided into groups that are essentially distinct from one another, with each group made up of individuals that resemble each other in equally intrinsic ways.

The main difficulty with attempting to order the diversity of the living world stems from the fact that organisms that resemble each other in some respects can be very different in others, and therefore any classification criteria must necessarily be arbitrary. Which characteristics should classifications be based upon? For example, is the presence of organs the most important physiological factor? Many water-dwelling creatures possess gills rather than lungs, for example, but this category includes so many organisms that it is of little use for classification purposes. This is especially true for cases such as the Lepidosiren (a genus of fish from the sub-class Dipneusti), which possess a double respiratory system with both lungs and gills. All vertebrates possess a heart, but this criterion alone cannot distinguish an elephant from a frog, and is not even unique to the class of vertebrates.

Earlier attempts at classification had therefore used only the criteria thought most suitable by the classifier for revealing the 'natural system'. This was frequently represented as the mysterious key used by the Creator to put order into nature, and the living world was often depicted as the 'great chain of being' of the philosopher Leibniz, which moved gradually from the simplest organisms to the most complex, from the 'inferior' to the 'superior'. From 1737 to 1738, the Swedish naturalist Carolus Linnaeus made his contribution to history not only by choosing the reproductive organs of plants as the criterion of his botanical classification but, most importantly, by inventing the clever system of binomial Latin nomenclature (one name for the genus, followed by a name for the species) which is still in use today.

As long as people believed, or tried to believe, that species were essentially invariable, the basic methodological confusion persisted. The more 'systems' accumulated and became ever more complex through the discovery of new species, the more the need for a different taxonomic approach became apparent to naturalists who had previously accepted the doctrine of fixed species. This occurred to Linnaeus himself at the end of his career, to Buffon in the volume *De l'âne* from his *Histoire naturelle*, and to Michel Adanson, who attempted to record every plant characteristic within what he called the 'natural method' in his *Familles naturelles des plantes* (1763) and came to a transformist conclusion. It was even more so the case with Lamarck, who supported transformism and rejected classification, despite having reclassified the flora of France early in his career. Most importantly, however, it was Darwin who, although

recognizing the practical convenience of classifications, went on to radically subvert them by putting all their divisions into perspective and treating them as symptoms of the unconscious assumption by classifiers themselves that genealogical order was the only valid natural order.

2. Minor characteristics

Darwin's first essay of 1842 contained some early critical reflections on classification, which he went on to expand in the second essay of 1844, and which were most completely expressed in *Origin of Species*. He concluded that it would be wrong for the most part to claim, as most of those who have devised systems of classification have done, that the physiological importance of an organ is the key criterion for classification purposes. In fact, he recognized, it is sometimes very minor physiological features that provide the clearest indications of a relationship between organisms.

'Numerous instances,' wrote Darwin, 'could be given of characters derived from parts which must be considered of very trifling physiological importance, but which are universally admitted as highly serviceable in the definition of whole groups. For instance, whether or not there is an open passage from the nostrils to the mouth, the only character, according to Owen, which absolutely distinguishes fishes and reptiles – the inflection of the angle of the lower jaw in marsupials – the manner in which the wings of insects are folded – mere colour in certain Algae – mere pubescence on parts of the flowers in grasses – the nature of the dermal covering, as hair or feathers, in the Vertebrata. If the Ornithorhynchus

had been covered with feathers instead of hair, this external and trifling character would have been considered by naturalists as an important aid in determining the degree of affinity of this strange creature to birds.'
(*Origin of Species*, chap. XIV).

3. Analogous or adaptive features

It is important to distinguish traits that are truly homologous from analogous or adaptive traits, which possess a superficial similarity due to the process of one organism's physical adaptation to an environment more usual to other organisms. The most frequently cited of these cases is the fishlike morphology of whales and other cetaceans, mammals which are completely adapted to underwater life.

Darwin showed that morphological similarities may often indicate a true affinity at the level of classification, but can never be universal. This is true in examples such as the dugong, a mammal of the order Sirenia, and the whale: or in the case of these two mammals and the fish; or of the mouse and the shrew; or the mouse and *Antechinus*, a small Australian marsupial. On the other hand, adaptive resemblances among the cetaceans themselves are true indication of a common ancestry. Ultimately, adaptive traits provide unmistakable proof both of the extraordinary capacity for variation among organisms (in the case of the cetaceans, a terrestrial mammal takes on the overall appearance of a fish) and, at the same time, of the extraordinary length of time that is required for this transformation. These are two essential ingredients for 'transformism'.

4. Putting static classifications and genealogies into perspective

Similar traits are only useful and relevant for classification purposes if the approach to classification itself is completely inverted, so that it is no longer similar traits that create classification groups, but the natural groups that cause these resemblances that become the clues to kinship.

It may be simple, Darwin explained, to determine a number of features that are common to all birds, but among the crustaceans it is extremely difficult to single out a common feature that connects two forms from the opposite extremes of the range. Nonetheless, between those two poles there is a spectrum of forms which share with their neighbours similarities so obvious that no member of the series could possibly be placed in another class. There are two phenomena that place all rules and customs of classification into perspective:

1. The instability of the categories of the classification itself (the addition of new species that display previously unobserved differences regularly disrupts the accepted hierarchy by, for example, raising a genus to the rank of sub-family or of family);

2. The extreme difficulty of many classifications, since naturalists often find it hard to decide whether a particular observed form is a species or a variety, for example.

Therefore, the key to ridding natural classification of its arbitrary nature is the one found in this hypothesis: living organisms are modified through descent combining inheritance and variation:

'All the foregoing rules and aids and difficulties in classification may be

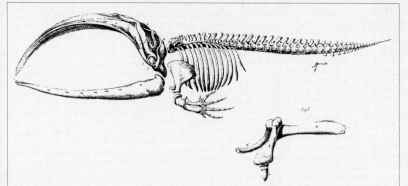

Side view of the skeleton of a right whale. Below right, an enlarged view of its rudimentary hind limb.

explained, if I do not greatly deceive myself, in the view that the Natural System is founded on descent with modification; that the characters which naturalists consider as showing true affinity between any two or more species, are those which have been inherited from a common parent, all true classification being genealogical; that community of descent is the hidden bond which naturalists have been unconsciously seeking, and not some unknown plan of creation.... Thus, the natural system is genealogical in its arrangement, like a pedigree.'

5. The principle of divergence

Varieties, species, genera, families, classes and all their subdivisions are merely a set of reference points on a branching genealogical diagram of evolutionary divergence. Darwin included this diagram (overleaf) in chapter IV of *Origin of Species*. ABCD EF GHIKL are the species of a major genus in a particular region. They are not equal in similarity to each other, hence the unequal spacing of the letters.

A is a common species, widely distributed, and varying a great deal. The branching lines represent its descent, which is also highly variable.

The space between two horizontal lines represents 1000 generations.

The italic lower-case letters, at the intersections with the horizontal lines, indicate marked varieties. For example, after 1000 generations, A has produced two marked varieties, a^1 and m^1. The farther one advances in this diagram, the more the varieties (modified descendants) differ amongst themselves and differ from their common ancestor. After 10,000 generations, species A has given rise to three forms, a^{10}, f^{10} and m^{10}, which are highly differentiated. If the extent of the modification between two horizontal lines is low, they will be considered marked varieties. But if the modification is extensive, accumulated or intense, they will be regarded as well-defined species. Varieties, therefore, are simply nascent species.

If the number of generations is increased still further, the top of the chart (simplified by Darwin) will show

Forelimb   Hind limb

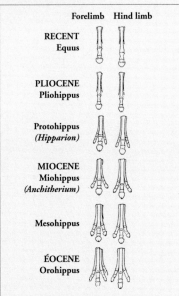

RECENT
Equus

PLIOCENE
Pliohippus

Protohippus
(Hipparion)

MIOCENE
Miohippus
(Anchitherium)

Mesohippus

ÉOCENE
Orohippus

From top to bottom, the evolution of the horse's hoof over the Tertiary era, drawn by O. C. Marsh, and showing the gradual reduction of the lateral digits in the fore and hind limbs.

eight species (from a^{14} to m^{14}) that are descended from A. When the same procedure is applied to species I, it produces two well-marked varieties or species (w^{10} and z^{10}) after 10,000 generations, and six new species (n^{14} to z^{14}) after 14,000 generations.

The branches that do not reach the top of the chart represent extinctions; sometimes these are rapid, and sometimes they occur after several marked varieties have been produced. The nine other species (BCD EF GHKL) have produced unmodified descendants over very variable lengths of time.

Overall, if very major modifications have occurred, species A will be replaced after 14,000 generations (or many more if one decides to accord a greater value to the intervals) by the eight species that have arisen from it (from a^{14} to m^{14}), and species I will be replaced by the six species that have derived from it (n^{14} to z^{14}). It can also be seen that, of the two species (E and F) which have not produced variations and which least resemble the species with a branching descent (A and I), and so consequently have suffered little from competition over habitats, only F has reached the 14,000th generation. 15 new species have taken the place of the 11 primitive species, only one of them without modification. The difference between a^{14} and z^{14} immediately appears to be greater, because of divergence, than that between the initial species.

The eight species that have arisen from A, and the six species that have arisen from I, regrouped at the top of the chart, are relatively close to each other within each group but very distinct from group to group, and could therefore form sub-genera, genera, or even sub-families, since all the intermediate species between A and I, with the exception of F, have become extinct over time.

What the diagram demonstrates is as follows: the more numerous a species is (= the more representatives it has), the more variable it is (= the more opportunities it offers for variation). The more variable it is, the more opportunities it offers natural selection (= the more opportunities it offers advantageous variations), and therefore, the more traits will diverge. The greater the divergence, the greater the possibilities for dominating habitats, and therefore the greater the chances for geographic expansion. Divergence in itself is an advantage.

C. EMBRYOLOGICAL EVIDENCE

'The weightiest of all to me,' as Darwin said in a letter to Lyell on 12 September 1860. He was naturally opposed to the idea of a separate successive creation of distinct types.

The development of some individual animals can be seen to involve a complex series of changes, such as the metamorphoses of insects and crustaceans, or the alternation of generations among some jellyfish. This means that an individual jellyfish, born of a fertilized egg, passes through an asexual polyp stage, which then becomes fixed and, through budding, produces detached discs which will become jellyfish, free and sexed, which themselves will go on to reproduce sexually. Before such transformation processes were understood, the separate stages involved were often believed by naturalists to be distinct beings, and were sometimes recorded as such by classification. Metamorphoses and alternation of generations allow for an overall view of the accelerated change from one form to another – from a caterpillar to a fully-formed butterfly or imago, or from a free-swimming larva to a fixed crustacean, or from a fixed polyp to a free-swimming jellyfish – and tend to contradict any belief in fixity in nature.

Comparison of the larval or embryonic stages and adult stages of many animals led Darwin to a similar conclusion. In 1828, the great Estonian embryologist Karl Ernst von Baer (1792–1876), the founder of modern

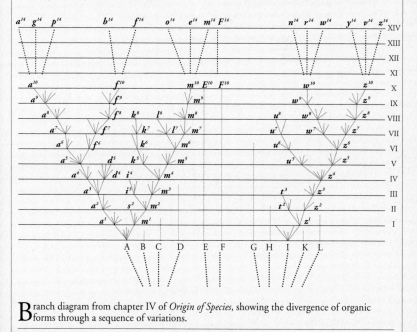

Branch diagram from chapter IV of *Origin of Species*, showing the divergence of organic forms through a sequence of variations.

embryology, set out the fundamental rules of animal development. These followed a path of differentiation, from the homogeneous to the heterogeneous, from the least differentiated to the most differentiated. Consequently, the first features to appear during embryonic development are the general characteristics of the type, then come those of the class, the order and other categories down to the species and the individual.

However, von Baer remained within a 'fixist' framework; like Cuvier, he distinguished four types or structural plans in organisms (Radiata, Articulata, Mollusca and Vertebrata) and rejected any transformist interpretation of his theory, opposing both the law of Meckel-Serres (which declared that the superior animals, during their embryonic development, pass through all the adult stages of the inferior animals) and also Darwin's more prudent but firmly transformist view of this 'law', which was later improved by Fritz Müller and systematized by the German naturalist Ernst Haeckel under the name of the 'fundamental biogenetic law' or the 'law of recapitulation': ontogeny recapitulates phylogeny. Von Baer believed that a vertebrate embryo, at any stage, remained a vertebrate embryo, and is not in any true sense the adult stage of a lesser form from which, in the course of its development, it grows increasingly distant. The only visible resemblances he believed in were those shared by different embryonic forms, and this is the belief that Darwin retained.

In chapter XIV of *Origin of Species*, Darwin discusses both the initial resemblance of various parts of the embryo that are later destined to produce clearly differentiated organs,

and also the similarities shared by the embryonic forms of animals which may be very different from each other as adults. This resemblance during the early stages of development cannot be due, in this last case, to shared environmental conditions: there is obviously no common ground between the young mammal carried and fed by its mother, the young bird enclosed in an egg incubated in a nest, and the spawn of a frog which develops in water.

Disregarding cases of precocious embryonic adaptation in some autonomous larvae – particularly insects – which are obliged to seek their own food, and thus to display a 'useful' differentiation at an early stage because of their living conditions and habits, it remains true that the examination of most larvae confirms the law of embryonic resemblance. Darwin knew through his own zoological experience how closely the larvae of pedunculated (stalked) cirripedes, such as goose barnacles, and sessile (stalkless) cirripedes, such acorn barnacles, resemble each other, even though they are very different in their adult form. He was also aware that many features visible in an embryo are and will remain useless, and that a few larvae, contrary to the rule, are more highly organized than their adult forms; for example, some cirripede larvae develop into 'complemental males' and are reduced to a sac containing the reproductive organs which lives on or inside the female.

'How, then,' asked Darwin, 'can we explain these several facts in embryology – namely, the very general, though not universal, difference in structure between the embryo and the adult; the various parts in the same individual embryo, which ultimately become very unlike and serve for diverse purposes, being at an early period of growth alike; the common, but not invariable, resemblance between embryos or larvae of the most distinct species in the same class; the embryo often retaining whilst within the egg or womb, structures which are of no service to it, either at that or at a later period of life; on the other hand larvae, which have to provide for their own wants, being perfectly adapted to the surrounding conditions; and lastly the fact of certain larvae standing higher in the scale of organization than the mature animal into which they are developed?'

Darwin answered this question by means of two observations that he generalized into principles:
• In the life of an individual, slight variations or individual differences do not appear at an early period. Two puppies, foals or young pigeons of very different breeds are less obviously different from each other than they will be as adults.
• Whatever the age at which a variation appears in the parent, it will tend to appear at the equivalent age in the offspring.

Together, these two principles (late variation and inheritance at corresponding ages) imply that, as a general rule, the young and, *a fortiori*, the early stages of embryonic development, are not obviously modified. In any species, forelimbs which were once legs in a distant ancestor, but in the course of evolution have become hands, flippers or wings, will begin to differ appreciably between the embryonic stage and the adult stage (the age when the variations useful to the animal for its own survival must have been selected and transmitted), while their embryonic stages will still

closely resemble each other. Likewise, if it was advantageous for a larva to modify to meet its own needs, these variations will have been selected and transmitted during the larval phase, and the rest of ontogenesis may therefore have taken a retrogressive path.

From these observations, it can be deduced that it is during the embryonic or larval stage of an animal that an approximate image of the adult ancestor of that group may most likely be found. Large and very different divisions of the class Crustacea share a larval form called *Nauplius*, which is not specifically adapted; this, therefore, provides an approximate image of this group's common ancestor. Similarly, comparison of the embryos of mammals, birds, reptiles and fish makes it clear that 'these animals are the modified descendants of some ancient progenitor, which was furnished in its adult state with branchiae, a swim-bladder, four fin-like limbs, and a long tail, all fitted for an aquatic life'.

Embryos are indicators of ancestry, and it is this line of descent which provides 'the hidden bond of connection which naturalists have been seeking under the term of the natural system'. This is retrospectively confirmed by the fact that many classifiers consider embryonic forms more valuable for taxonomic purposes than adult forms. Common embryonic or larval conformation is evidence for a common origin, while developmental and adult differences, caused by adaptation to new living conditions, occur later and obscure these similarities. It was examination of larvae that made it possible to remove the cirripedes from their incorrect classification as molluscs, and to classify them correctly as crustaceans.

Fish (Bony)	Salamander (Amphibian)	Hen (Bird)	Man (Mammal)

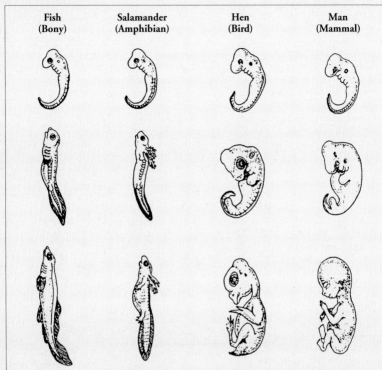

This table, taken from the work of Karl Ernst von Baer, shows three successive stages of embryonic development and illustrates the differentiation processes that eventually result in distinct and differentiated species. In their earliest phases, the embryos of these four species, from the classes of bony fish, amphibians, birds and mammals, closely resemble each other. Evolutionary theory uses this close resemblance at the early stages of ontogenesis as evidence of the common ancestry of all living things.

For the purposes of 'transformism', similarity is usually interpreted as evidence for kinship, but similarity early in life is, even more importantly, evidence for a common origin – evidence that is not yet blurred by convergent phenomena (for example among the cetaceans) or by the suppression of developmental stages – processes linked to different adaptive needs. Within a class, embryonic similarity provides a basic template that Darwin described as 'a picture, more or less obscured, of the common parent-form of each great class of animals'.

Patrick Tort

Contemporaries, critics and friends

Darwin's theory brought worldwide recognition to the principle of the transformation of species. Nonetheless, the theory of evolution could be understood in several ways, and the most famous Darwinists were not necessarily the most Darwinian in outlook.

T. H. Huxley (1825–95)

Thomas Henry Huxley, nicknamed 'Darwin's bulldog', was a talented anatomist and deeply committed to the freedom of science. The sworn enemy of Richard Owen, he struggled throughout his career to reach a position of power within the English scientific establishment. After reading Origin of Species *in 1859, he began to champion natural selection, although he did have some criticisms of the theory, favouring evolution by abrupt leaps over the Darwinian accumulation of slight variations. His best-known book,* Evidence as to Man's Place in Nature *(1863), highlights the morphological and anatomical relationship between humans and the anthropoid apes. His confrontation with Samuel Wilberforce during the famous Oxford Meeting of 1860 brought him lasting fame. Huxley eventually became president of the Royal Society.*

It cannot be doubted, I think, that Mr Darwin has satisfactorily proved that what he terms selection, or selective modification, must occur, and does occur, in nature; and he has also proved to superfluity that such selection is competent to produce forms as distinct, structurally, as some genera even are. If the animated world presented us with none but structural differences, I should have no hesitation in saying that Mr Darwin had demonstrated the existence of a true physical cause, amply competent to account for the origin of living species, and of man among the rest.

But, in addition to their structural distinctions, the species of animals and plants, or at least a great number of them, exhibit physiological characters – what are known as distinct species, structurally, being for the most part either altogether incompetent to breed one with another; or if they breed, the resulting mule, or hybrid, is unable to perpetuate its race with another hybrid of the same kind.

A true physical cause is, however, admitted to be such only on one condition – that it shall account for all the phenomena which come within the range of its operation. If it is inconsistent with any one phenomenon, it must be rejected; if it fails to explain any one phenomenon, it is so far weak, so far to be suspected; though it may have a perfect right to claim provisional acceptance.

Now, Mr Darwin's hypothesis is not, so far as I am aware, inconsistent with any known biological fact; on the contrary, if admitted, the facts of Development, of Comparative

Anatomy, of Geographical Distribution, and of Palaeontology, become connected together, and exhibit a meaning such as they never possessed before; and I, for one, am fully convinced, that if not precisely true, that hypothesis is as near an approximation to the truth as, for example, the Copernican hypothesis was to the true theory of the planetary motions.

But, for all this, our acceptance of the Darwinian hypothesis must be provisional so long as one link in the chain of evidence is wanting; and so long as all the animals and plants certainly produced by selective breeding from a common stock are fertile, and their progeny are fertile with one another, that link will be wanting. For, so long, selective breeding will not be proved to be competent to do all that is required of it to produce natural species.

I have put this conclusion as strongly as possible before the reader, because the last position in which I wish to find myself is that of an advocate for Mr Darwin's, or any other views; if by an advocate is meant one whose business it is to smooth over real difficulties, and to persuade where he cannot convince.

In justice to Mr Darwin, however, it must be admitted that the conditions of fertility and sterility are very ill understood, and that every day's advance in knowledge leads us to regard the hiatus in his evidence as of less and less importance, when set against the multitude of facts which harmonize with, or receive an explanation from, his doctrines.

I adopt Mr Darwin's hypothesis, therefore, subject to the production of proof that physiological species may be produced by selective breeding; just as a physical philosopher may accept the undulatory theory of light, subject to the proof of the existence of the hypothetical ether; or as the chemist adopts the atomic theory, subject to the proof of the existence of atoms; and for exactly the same reasons, namely, that it has an immense amount of prima facie probability: that it is the only means at present within reach of reducing the chaos of observed facts to order; and lastly, that it is the most powerful instrument of investigation which has been presented to naturalists since the invention of the natural system of classification, and the commencement of the systematic study of embryology.

T. H. Huxley
Evidence as to Man's Place in Nature,
1863

Herbert Spencer (1820–1903)

An engineer and philosopher, Spencer was a Lamarckist and based his 'synthetic system of philosophy' on sociological and ethical principles which tended to confuse Darwinism and social selection. Seeing society as one enormous organism, like Lamarck he believed in the direct action of the environment, and only used natural selection to justify competition between individuals, resulting in the elimination of the 'least fit'. He became the father of 'Social Darwinism', based on the idea that natural selection should be allowed free rein within society, and that charitable action to aid the 'natural' victims of elimination should be avoided. Fiercely opposed to the authority of the state and in favour of absolute freedom for the individual, he was the first to systematize what would later be called sociobiology. Darwin borrowed Spencer's term 'the survival of the fittest', although in his Autobiography *he notes his personal dislike of Spencer and the great differences in their approaches to science.*

Already we have seen that in so far as the members of a species are subject to different sets of incident forces, they are differentiated, or divided into varieties. Here it remains to add that such of them as are subject to like sets of incident forces, are segregated. For by the process of 'natural selection', there is a continual purification of each species from those individuals which depart from the common type in ways that unfit them for the conditions of their existence. Consequently, there is a continual leaving behind of those individuals which are in all respects fit for the conditions of their existence, and are therefore nearly alike. The circumstances to which any species is exposed, being an involved combination of incident forces; and the members of the species having among them some that differ more than is usual from the

average structure required for meeting these forces; it results that these forces are constantly separating such divergent individuals from the rest, and so preserving the uniformity of the rest – keeping up its integrity as a species or variety. Just as the changing autumn leaves are picked out by the wind from among the green ones around them, or just as, to use Prof. Huxley's simile, the smaller fragments pass through a sieve while the larger are kept back; so, the uniform incidence of external forces affects the members of a group of organisms similarly in proportion as they are similar, and differently in proportion as they are different; and thus is ever segregating the like by parting the unlike from them. Whether these separated members are killed off, as mostly happens, or whether, as otherwise happens, they survive and multiply into a distinct variety, in consequence of their fitness to certain partially-unlike conditions, matters not to the argument. The one case conforms to the law that the unlike units of an aggregate are sorted into their kinds and parted, when uniformly subject to the same incident forces, and the other to the converse law that the like units of an aggregate are parted and separately grouped when subject to different incident forces. And on consulting Mr Darwin's remarks on divergence of character, it will be seen that the segregations thus caused tend ever to become more definite.

Herbert Spencer
First Principles, 1862

Samuel Wilberforce (1805–73)

Son of the anti-slavery campaigner William Wilberforce, 'Soapy Sam', as he was known, was Bishop of Oxford and a charismatic and eloquent public speaker, as well as a fervent believer in the doctrine of natural theology. This extract is taken from a critical review of Origin of Species *that he wrote shortly before the famous Oxford Meeting at which he publically attacked Darwin's theory.*

Any contribution to our Natural History literature from the pen of Mr C. Darwin is certain to command attention. His scientific attainments, his insight and carefulness as an observer, blended with no scanty measure of imaginative sagacity, and his clear and lively style, make all his writings unusually attractive. His present volume on the 'Origin of Species' is the result of many years of observation, thought, and speculation; and is manifestly regarded by him as the 'opus' upon which his future fame is to rest. It is true that he announces it modestly enough as the mere precursor of a mightier volume. But that volume is only intended to supply the facts which are to support the complete argument of the present essay. In this we have a specimen-

collection of the vast accumulation; and, working from these as the high analytical mathematician may work from the admitted results of his conic sections, he proceeds to deduce all the conclusions to which he wishes to conduct his readers.

The essay is full of Mr Darwin's characteristic excellences. It is a most readable book; full of facts in natural history, old and new, of his collecting and of his observing; and all of these are told in his own perspicuous language, and all thrown into picturesque combinations, and all sparkle with the colours of fancy and the lights of imagination. It assumes, too, the grave proportions of a sustained argument upon a matter of the deepest interest, not to naturalists only, or even to men of science exclusively, but to every one who is interested in the history of man and of the relations of nature around him to the history and plan of creation. With Mr Darwin's 'argument' we may say in the outset that we shall have much and grave fault to find...

Samuel Wilberforce
The Quarterly Review, vol. 108, 1860.

Joseph Dalton Hooker (1817–1911)

A talented botanist with a particular interest in plant geography, Hooker was a close friend of Darwin's for many years and a fervent supporter of the theory of natural selection. Together with Charles Lyell, he presented the theories of both Darwin and Alfred Russel Wallace to the Linnean Society in 1858, and he also participated in the famous Oxford Meeting, a confrontation between Samuel Wilberforce and T. H. Huxley over Origin of Species. *In a letter to Darwin, Hooker recounts the events in Oxford in his own words.*

... Huxley & Owen had had a furious battle over Darwin's absent body at Section D., before my arrival.... H. was triumphant – You and your book forthwith became the topics of the day... however, hearing that Soapy Sam was to answer I waited to hear the end. The meeting was so large that they had adjourned to the Library which was crammed with between 700 & 1000 people, for all the world was there to hear Sam Oxon – Well Sam Oxon got up & spouted for half an hour with inimitable spirit ugliness & emptiness & unfairness, I saw he was coached up by Owen & knew nothing & he said not a syllable but what was in the Reviews – he ridiculed you badly & Huxley savagely – Huxley answered admirably & turned the tables, but he could not throw his voice over so large an assembly, nor command the audience; & he did not allude to Sam's weak points nor put the matter in a form or way that carried the audience. The battle waxed hot. Lady Brewster fainted, the excitement increased as others spoke – my blood boiled, I felt myself a dastard; now I saw my advantage – I swore to myself I would smite that Amalekite Sam hip & thigh if my heart jumped out of my mouth & I handed my name up to the President (Henslow) as ready to throw down the gauntlet – I must tell you that Henslow as president would have none speak but those who had *arguments* to use, & 4 persons had been burked by the audience & President for mere declamation: it moreover became necessary for each speaker to mount the platform & so there I was cocked up with Sam at my right elbow, & there & then I smashed him amid rounds of applause – I hit him in the wind at the first shot in 10 words taken from his own ugly mouth – & then proceeded to demonstrate in as few more, 1) that he could never have read your book & 2) that he was absolutely ignorant of the rudiments of Bot. Science – I said a few more on the subject of my own experience, & conversion & would up with a very few observations on the relative position of the old & new hypotheses & with some words of caution to the audience – Sam was shut up – had not one word to say in reply & the meeting was dissolved forthwith leaving you master of the field after 4 hours battle. Huxley who had borne all the previous brunt of the battle & who had never before (thank God) praised me to my face, told me it was splendid, & that he did not know before what stuff I was made of – I have been congratulated & thanked by the blackest coats & whitest stocks in Oxford (for they hate their Bishop...) & plenty of ladies too have flattered me....

J. D. Hooker, 2 July 1860

Ernst Haeckel (1834–1919)

German zoologist, professor at the University of Jena (1862–68). A committed 'transformist', chiefly inspired by Lamarck, his talent for synthesis and popularization of new ideas made him highly influential. He was a skilled naturalist and draughtsman, a passionate polemicist and propagandist, and a staunch opponent of the Pope and religion in general. He popularized the 'fundamental biogenetic law', a theory that the successive stages of development (ontogeny) of the higher vertebrates recapitulate the adult stages of their ancestors (phylogeny) within an evolution to which he pictured as a branching genealogical tree. He was an expert in

marine biology, a specialist in unicellular organisms (protists), a brilliant morphologist (Generelle Morphologie der Organismen, 1866) and a skilled embryologist, although his diagrams exaggerated the similarities between vertebrate embryos. Unfortunately, he also supported the conservative rule of Bismarck, pioneered 'social Darwinism' in Germany, and propagated eugenic concepts such as 'Spartan selection', which were later catastrophically taken up by the Third Reich.

In the short time that has passed since the appearance of Charles Darwin's book *On the Origin of Species in the Animal and Vegetable Kingdom*, the History of Evolution has advanced so greatly that it is scarcely possible to point to an equally great advance throughout the whole record of the Natural Sciences. The literature of Darwinism is increasing day by day, not only in connection with Zoology and Botany – which are the specian sciences most affected and reformed by the Darwinian Theory – but far beyond. It is applied in much wider circles with a zeal and interest which no other scientific theory has ever aroused....

While Lamarck explained the variation of organisms descended from common ancestral forms, as especially the effect of habit and the use of the organs, but also by the aid of the phenomena of Heredity, Darwin independently, and on an entirely new basis, unfolded the actual causes which mechanically accomplish the modification of the various animal and vegetable forms by the aid of Adaptation and Heredity. Darwin deduced his 'Theory of Selection' from the following considerations.

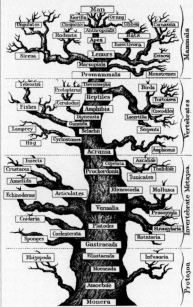

Haeckel's tree of life, taken from *The Evolution of Man (Anthropogenie)*, 1879. At the base of the tree are the Monera, the simplest single-celled creatures, mucous globules of homogenous, unstructured plasma without a nucleus or a fixed shape. Haeckel wrongly believed that these microscopic organisms were spontaneously generated and occupied a transitional zone between organic and inorganic matter.

He compared the origin of the various breeds of animals and plants which man is able to produce artificially, – the conditions of 'Selection' in horticulture, and in the breeding of domestic animals, – with the origin of wild species of plants and animals in a natural state. He thus found that causes similar to those which, in artificially breeding domestic animals, and raising cultivated plants, we apply to alter the forms, are also at work in Nature. He named the most effective of all the co-operating causes the Struggle for Existence. The gist of Darwin's theory, properly so called, is this simple idea: that the Struggle for Existence in Nature evolves new Species without design, just as the Will of Man produces new Varieties in Cultivation with design. Just as the gardener and the farmer breed for their own advantage, and according to their own will, making judicious use of the productive effects of Heredity and Adaptation, so does the Struggle for Existence constantly modify the forms of vegetables and animals in an undomesticated state. This Struggle for Existence, or the universal efforts of organisms to secure the necessary means of existence, works without design, but yet in the same way modifies the organisms. But as under its influence Heredity and Adaptation enter into most intimate reciprocal relations, there necessarily arise new forms, or variations, which are of advantage to the organism, and which have, therefore, an object, although in reality not originating from a preconceived design.

This simple fundamental idea is the real gist of Darwinism, or the 'Theory of Selection'. Its author conceived the idea long ago, but with admirable industry he employed twenty years in collecting data from actual experience for proving his theory before declaring it.

Ernst Haeckel
The Evolution of Man, 1879

Asa Gray (1810–88)

An American botanist who became a friend and favourite correspondent of Darwin. Despite his Christian beliefs and adherence to the principles of natural theology, Gray was primarily responsible for introducing Darwin's ideas to America. A great classifier, collector and specialist in North American flora, he taught at Harvard and became an opponent of the famous Louis Agassiz. In 1859 he espoused the theory of natural selection, which he attempted to reconcile with his faith. Darwin did not agree with Gray's finalist beliefs but nonetheless encouraged him to promote the theory of natural selection in the United States.

This book is already exciting much attention. Two American editions are

announced, through which it will become familiar to many of our readers, before these pages are issued. An abstract of the argument – for 'the whole volume is one long argument', as the author states – is unnecessary in such a case; and it would be difficult to give by detached extracts. For the volume itself is an abstract, a prodromus of a detailed work upon which the author has been laboring for twenty years, and which 'will take two or three more years to complete'. It is exceedingly compact; and although useful summaries are appended to the several chapters, and a general recapitulation contains the essence of the whole, yet much of the aroma escapes in the treble distillation, or is so concentrated that the flavor is lost to the general or even to the scientific reader. The volume itself – the proof-spirit – is just condensed enough for its purpose. It will be far more widely read, and perhaps will make deeper impression, than the elaborate work might have done, with all its full details of the facts upon which the author's sweeping conclusions have been grounded. At least it is a more readable book: but all the facts that can be mustered in favor of the theory are still likely to be needed.

Who, upon a single perusal, shall pass judgment upon a work like this, to which twenty of the best years of the life of a most able naturalist have been devoted? And who among those naturalists who hold a position that entitles them to pronounce summarily upon the subject, can be expected to divest himself for the nonce of the influence of received and favorite systems? In fact, the controversy now opened is not likely to be settled in an off-hand way, nor is it desirable that it should be. A spirited conflict among

opinions of every grade must ensue, which – to borrow an illustration from the doctrine of the book before us – may be likened to the conflict in Nature among races in the struggle for life, which Mr Darwin describes; through which the views most favored by facts will be developed and tested by 'Natural Selection', the weaker ones be destroyed in the process, and the strongest in the long-run alone survive.

Asa Gray
*Darwiniana: Essays and Reviews
pertaining to Darwinism*, 1876

'Monkeyana'

Controversy over evolution provided prime material for satirists of the day. The many cartoons and parodies published included this poem from Punch, *May 1861.*

Am I satyr or man?
Pray tell me who can,
And settle my place in the scale.
A man in ape's shape,
An anthropoid ape,
Or monkey deprived of his tail?

The Vestiges taught,
That all came from naught
By 'development', so called, 'progressive';
That insects and worms
Assume higher forms
By modification excessive.

Then Darwin set forth
In a book of much worth,
The importance of 'nature's selection';
How the struggle for life
Is a laudable strife,
And results in 'specific distinction'.

Let pigeons and doves
Select their own loves,
And grant them a million of ages,

Then doubtless you'll find
They've altered their kind,
And changed into prophets and sages.

Leonard Horner relates,
That Biblical dates
The age of the world cannot trace;
That Bible tradition,
By Nile's deposition,
Is put to the right about face.

Then there's Pengelly
Who next will tell ye
That he and his colleagues of late
Find celts and shaped stones
Mixed up with cave bones
Of contemporaneous date.

Then Prestwich, he pelts
With hammers and celts
All who do not believe his relation,
That the tools he exhumes
From gravelly tombs
Date before the Mosaic creation.

Then Huxley and Owen,
With rivalry glowing,
With pen and ink rush to the scratch;
'Tis Brain versus Brain,
Till one of them's slain,
By Jove! it will be a good match!

Says Owen, you can see
The brain of Chimpanzee
Is always exceedingly small,
With the hindermost 'horn'
Of extremity shorn,
And no 'Hippocampus' at all.

The Professor then tells 'em,
That man's 'cerebellum',
From a vertical point you can't see;
That each 'convolution'
Contains a solution
Of 'Archencephalic' degree.

That apes have no nose,
And thumbs for great toes,
And a pelvis both narrow and slight;
They can't stand upright,
Unless to show fight,
With 'Du Chaillu', that chivalrous knight!

Next Huxley replies,
That Owen he lies,
And garbles his Latin quotation;
That his facts are not new,
His mistakes not a few,
Detrimental to his reputation.

'To twice slay the slain,
By dint of the Brain,
(Thus Huxley concludes his review)
Is but labour in vain,
Unproductive of gain,
And so I shall bid you Adieu!'

'GORILLA'
Zoological Gardens, May 1961

THE HORNET.

A VENERABLE ORANG-OUTANG.
A CONTRIBUTION TO UNNATURAL HISTORY.

The Scopes 'Monkey Trial'

Darwin's theory of evolution has undergone regular attacks by religious fundamentalists. One attack in 1925 resulted in one of the most famous trials of the twentieth century. In Dayton, Tennessee, high-school teacher John Scopes was tried for teaching evolution in biology lessons.

The prosecution was really little more than a stunt, conceived with the aim of making a stand against recently introduced anti-evolution legislation in the state of Tennessee, with the welcome side effect of bringing business to the town. The storm of publicity surrounding the case climaxed in a public confrontation between crusading fundamentalist William Jennings Bryan for the prosecution and liberal, agnostic lawyer Clarence Darrow for the defence. The following are extracts from the court transcripts of the day.

The testimony of Howard Morgan, a 14-year-old student of Scopes:

General Stewart for the prosecution: Now, you say you were studying this book in April; how did Prof. Scopes teach that book to you? I mean by that did he ask them questions and you answered them or did he give you lectures, or both? Just explain to the jury here now, these gentlemen here in front of you, how he taught the books to you.

A: Well, sometimes he would ask us questions and then he would lecture to us on different subjects in the book.

Q: Sometimes he asked you questions and sometimes lectured to you on different subjects in the book?

A: Yes, sir.

Q: Did he ever undertake to teach you anything about evolution?

A: Yes, sir....

Q: Just state in your own words, Howard, what he taught you and when it was.

A: It was along about the 2nd of April.

Q: Of this year?

A: Yes, sir; of this year. He said that the earth was once a hot molten mass too hot for plant or animal life to exist upon it; in the sea the earth cooled off; there was a little germ of one cell organism formed, and this organism kept evolving until it got to be a pretty good-sized animal, and then came on to be a land animal and it kept on evolving, and from this was man.

Q: Let me repeat that; perhaps a little stronger than you. If I don't get it right, you correct me.

Hays (defending): Go to the head of the class....

Stewart: I ask you further, Howard, how did he classify man with reference to other animals; what did he say about them?

A: Well, the book and he both classified man along with cats and dogs, cows, horses, monkeys, lions, horses and all that.

Q: What did he say they were?

A: Mammals.

Q: Classified them along with dogs, cats, horses, monkeys and cows?
A: Yes, sir.

Cross examination by Mr. Darrow:
Q: Let's see, your name is what?
A: Howard Morgan.
Q: Now, Howard, what do you mean by classify?
A: Well, it means classify these animals we mentioned, that men were just the same as them, in other words.
Q: He didn't say a cat was the same as a man?
A: No, sir: he said man had a reasoning power; that these animals did not.
Q: There is some doubt about that, but that is what he said, is it? (Laughter in the courtroom.)
The Court: Order.
Stewart: With some men.
Darrow: A great many.
Q: Now, Howard, he said they were all mammals, didn't he?
A: Yes, sir.
Q: Did he tell you what a mammal was, or don't you remember?
A: Well, he just said these animals were mammals and man was a mammal.
Q: No; but did he tell you what distinguished mammals from other animals?
A: I don't remember.
Q: If he did, you have forgotten it? Didn't he say that mammals were those beings which suckled their young?
A: I don't remember about that.
Q: You don't remember?
A: No.
Q: Do you remember what he said that made any animal a mammal, what it was or don't you remember?
A: I don't remember.
Q: But he said that all of them were mammals?
A: All what?

Q: Dogs and horses, monkeys, cows, man, whales, I cannot state all of them, but he said all of those were mammals?
A: Yes, sir; but I don't know about the whales; he said all those other ones. (Laughter in the courtroom.)
The Court: Order....
Q: Well, did he tell you anything else that was wicked?
A: No, not that I remember of....
Q: Now, he said the earth was once a molten mass of liquid, didn't he?
A: Yes.
Q: By molten, you understand melted?
A: Yes, sir.
Q: After that, it got cooled enough and the soil came, that plants grew; is that right?
A: Yes, sir, yes, sir.
Q: And that the first life was in the sea.
A: Yes, sir.
Q: And that it developed into life on the land?
A: Yes, sir.
Q: And finally into the highest organism which is known to man?
A: Yes, sir.
Q: Now, that is about what he taught you? It has not hurt you any, has it?
A: No, sir.
Darrow: That's all.

Later in the trial, the defence counsel Darrow made the highly theatrical move of calling the prosecuting counsel Bryan to the stand as an 'expert' on the Biblical account of the Creation. Bryan agreed to testify.

Darrow: You believe the story of the flood to be a literal interpretation?
Bryan: Yes, sir.
Q: When was that Flood?
A: I would not attempt to fix the date. The date is fixed, as suggested this morning.

Q: About 4004 B.C.?

A: That has been the estimate of a man that is accepted today. I would not say it is accurate.

Q: That estimate is printed in the Bible?

A: Everybody knows, at least, I think most of the people know, that was the estimate given.

Q: But what do you think that the Bible itself says? Don't you know how it was arrived at?

A: I never made a calculation.

Q: A calculation from what?

A: I could not say.

Q: From the generations of man?

A: I would not want to say that.

Q: What do you think?

A: I do not think about things I don't think about.

Q: Do you think about things you do think about?

A: Well, sometimes.

(Laughter in the courtyard.)

Policeman: Let us have order....

Stewart: Your honor, he is perfectly able to take care of this, but we are attaining no evidence. This is not competent evidence.

Bryan: These gentlemen have not had much chance: they did not come here to try this case. They came here to try revealed religion. I am here to defend it and they can ask me any question they please.

The Court: All right.

(Applause from the court yard.)

Darrow: Great applause from the bleachers.

Bryan: From those whom you call 'yokels'.

Q: I have never called them yokels.

A: That is the ignorance of Tennessee, the bigotry.

Q: You mean who are applauding you? (Applause.)

A: Those are the people whom you insult.

Q: You insult every man of science and learning in the world because he does not believe in your fool religion.

The Court: I will not stand for that.

Darrow: For what he is doing?

The Court: I am talking to both of you....

Q: Wait until you get to me. Do you know anything about how many people there were in Egypt 3,500 years ago, or how many people there were in China 5,000 years ago?

A: No.

Q: Have you ever tried to find out?

A: No, sir. You are the first man I ever heard of who has been in interested in it. (Laughter.)

Q: Mr. Bryan, am I the first man you ever heard of who has been interested in the age of human societies and primitive man?

A: You are the first man I ever heard speak of the number of people at those different periods.

Q: Where have you lived all your life?

A: Not near you. (Laughter and applause.)

Q: Nor near anybody of learning?

A: Oh, don't assume you know it all.

Q: Do you know there are thousands of books in our libraries on all those subjects I have been asking you about?

A: I couldn't say, but I will take your word for it....

Q: Have you any idea how old the earth is?

A: No.

Q: The book you have introduced in evidence tells you, doesn't it?

A: I don't think it does, Mr. Darrow.

Q: Let's see whether it does; is this the one?

A: That is the one, I think.

Q: It says B.C. 4004?

A: That is Bishop Usher's calculation.

Q: That is printed in the Bible you introduced?

A: Yes, sir....

Q: Would you say that the earth was only 4,000 years old?

A: Oh, no; I think it is much older than that.

Q: How much?

A: I couldn't say.

Q: Do you say whether the Bible itself says it is older than that?

A: I don't think it is older or not.

Q: Do you think the earth was made in six days?

A: Not six days of twenty-four hours.

Q: Doesn't it say so?

A: No, sir....

The Court: Are you about through, Mr. Darrow?

Darrow: I want to ask a few more questions about the creation.

The Court: I know. We are going to adjourn when Mr. Bryan comes off the stand for the day. Be very brief, Mr. Darrow. Of course, I believe I will make myself clearer. Of course, it is incompetent testimony before the jury. The only reason I am allowing this to go in at all is that they may have it in the appellate court as showing what the affidavit would be.

Bryan: The reason I am answering is not for the benefit of the superior court. It is to keep these gentlemen from saying I was afraid to meet them and let them question me, and I want the Christian world to know that any atheist, agnostic, unbeliever, can question me anytime as to my belief in God, and I will answer him.

Darrow: I want to take an exception to this conduct of this witness. He may be very popular down here in the hills....

Bryan: Your honor, they have not asked a question legally and the only reason they have asked any question is for the purpose, as the question about Jonah was asked, for a chance to give this agnostic an opportunity to criticize a believer in the world of God; and I answered the question in order to shut his mouth so that he cannot go out and tell his atheistic friends that I would not answer his questions. That is the only reason, no more reason in the world.

Malone (defending): Your honor on this very subject, I would like to say that I would have asked Mr. Bryan: and I consider myself as good a Christian as he is: every question that Mr. Darrow has asked him for the purpose of bring out whether or not there is to be taken in this court a literal interpretation of the Bible, or whether, obviously, as these questions indicate, if a general and literal construction cannot be put upon the parts of the Bible which have been covered by Mr. Darrow's questions. I hope for the last time no further attempt will be made by counsel on the other side of the case, or Mr. Bryan, to say the defense is concerned at all with Mr. Darrow's particular religious views or lack of religious views. We are here as lawyers with the same right to our views. I have the same right to mine as a Christian as Mr. Bryan has to his, and we do not intend to have this case charged by Mr. Darrow's agnosticism or Mr. Bryan's brand of Christianity. (A great applause.)

Darrow: Mr. Bryan, do you believe that the first woman was Eve?

A: Yes.

Q: Do you believe she was literally made out of Adams's rib?

A: I do.

Q: Did you ever discover where Cain got his wife?

A: No, sir; I leave the agnostics to hunt for her.

Q: You have never found out?

A: I have never tried to find.

Q: You have never tried to find?

A: No.

Q: The Bible says he got one, doesn't it? Were there other people on the earth at that time?

A: I cannot say.

Q: You cannot say. Did that ever enter your consideration?

A: Never bothered me.

Q: There were no others recorded, but Cain got a wife.

A: That is what the Bible says.

Q: Where she came from you do not know. All right. Does the statement, 'The morning and the evening were the first day', and 'The morning and the evening were the second day', mean anything to you?

A: I do not think it necessarily means a twenty-four-hour day.

Q: You do not?

A: No.

Q: What do you consider it to be?

A: I have not attempted to explain it. If you will take the second chapter: let me have the book. (Examining Bible.) The fourth verse of the second chapter says: 'These are the generations of the heavens and of the earth, when they were created in the day that the Lord God made the earth and the heavens', the word 'day' there in the very next chapter is used to describe a period. I do not see that there is any necessity for construing the words, 'the evening and the morning', as meaning necessarily a twenty-four-hour day, 'in the day when the Lord made the heaven and the earth'.

Q: Then, when the Bible said, for instance, 'and God called the firmament heaven. And the evening and the morning were the second day', that does not necessarily mean twenty-four hours?

A: I do not think it necessarily does.

Q: Do you think it does or does not?

A: I know a great many think so.

Q: What do you think?

A: I do not think it does.

Q: You think those were not literal days?

A: I do not think they were twenty-four-hour days.

Q: What do you think about it?

A: That is my opinion: I do not know that my opinion is better on that subject than those who think it does.

Q: You do not think that?

A: No. But I think it would be just as easy for the kind of God we believe in to make the earth in six days as in six years or in 6,000,000 years or in 600,000,000 years. I do not think it important whether we believe one or the other.

Q: Do you think those were literal days?

A: My impression is they were periods, but I would not attempt to argue as against anybody who wanted to believe in literal days.

Q: I will read it to you from the Bible: 'And the Lord God said unto the serpent, because thou hast done this, thou art cursed above all cattle, and above every beast of the field; upon thy belly shalt thou go and dust shalt thou eat all the days of thy life.' Do you think that is why the serpent is compelled to crawl upon its belly?

A: I believe that.

Q: Have you any idea how the snake went before that time?

A: No, sir.

Q: Do you know whether he walked on his tail or not?

A: No, sir. I have no way to know. (Laughter in audience.)

Q: Now, you refer to the cloud that was put in heaven after the flood, the rainbow. Do you believe in that?

A: Read it.

Q: All right, Mr. Bryan, I will read it for you.

Bryan: Your Honor, I think I can shorten this testimony. The only purpose Mr. Darrow has is to slur at the Bible, but I will answer his question. I will answer it all at once, and I have no objection in the world, I want the world to know that this man, who does not believe in a God, is trying to use a court in Tennessee...

Darrow: I object to that.

Bryan: (continuing) ... to slur at it, and while it will require time, I am willing to take it.

Darrow: I object to your statement. I am exempting you on your fool ideas that no intelligent Christian on earth believes.

The Court: Court is adjourned until nine o'clock tomorrow morning.

At the end of the trial, the defence asked the jury to return a verdict of 'guilty', so that the case could be appealed to the Tennessee Supreme Court as a challenge to the anti-evolution statute itself.

Court: Mr. Foreman, will you tell us whether you have agreed on a verdict?

Foreman: Yes, sir, we have your honor.

Court: What do you find?

Foreman: We have found for the state, found the defendant guilty.

Court: Did you fix the fine?

Foreman: No, sir.

Court: You leave it to the court?

Foreman: Leave it to the court.

Court: Mr. Scopes, will you come around here, please, sir.

(The defendant presents himself before the court.)

Court: Mr. Scopes, the jury has found you guilty under this indictment, charging you with having taught in the schools of Rhea county, in violation of what is commonly known as the anti-evolution statute, which makes it unlawful for any teacher to teach in any of the public schools of the state, supported in whole or in part by the public school funds of the state, any theory that denies the story of the divine creation of man, and teach instead thereof that man has descended from a lower order of animals. The jury have found you guilty. The statute makes this an offense punishable by fine of not less than $100 nor more than $500. The court now fixes your fine at $100, and imposes that fine upon you. Have you anything to say, Mr. Scopes, as to why the court should not impose punishment upon you?

Defendant J. T. Scopes: Your honor, I feel that I have been convicted of violating an unjust statute. I will continue in the future, as I have in the past, to oppose this law in any way I can. Any other action would be in violation of my ideal of academic freedom – that is, to teach the truth as guaranteed in our constitution of personal and religious freedom. I think the fine is unjust.

Dayton, Tennessee, 1925

The voyage of the *Beagle*

A brief chronology of Darwin's travels

1831

27 **December,** departure from Devonport, England.

1832

16 **January–8 February**, Cape Verde [volcanic islands].
15–16 **February,** rocks of São Paulo [volcanic islands].
20 **February,** Fernando de Noronha [volcanic island].
29 **February–18 March,** Bahia, Brazil.
4 **April–5 July,** *Beagle* anchors at Rio de Janeiro.
8–23 **April**, overland journey [witnesses and condemns slavery].
26 **July–19 August,** Montevideo, Uruguay.
6 **September–17 October,** Bahía Blanca, Argentina [discovers a *Megatherium* jawbone at Punta Alta. Records the site].
2–10 **November,** Buenos Aires.
14–27 **November,** Montevideo [receives volume II of Lyell's *Principles of Geology*].
16 **December–26 February 1833,** Tierra del Fuego.

Fossil fragments from *Scelidotherium* and *Mylodon*

1833

1 **March–6 April,** Falkland Islands.
28 **April–23 July,** Maldonado, Uruguay.
3–24 **August,** mouth of the Rio Negro, Argentina.
11–17 **August,** travels from El Carmen to Bahía Blanca.
24 **August–6 October,** Survey of the Argentinian coast.
31 **August,** excavations at Punta Alta.
8–20 **September,** overland trip from Bahía Blanca to Buenos Aires.
6–19 **October,** Maldonado.
27 **September–20 October,** overland trip from Buenos Aires to Santa Fé, returning along the Paraná river. Stays five days at Santa Fé Bajada; excavates fossilized animal remains (armadillo, toxodon, mastodon, horse) and returns downriver.
21 **October–6 December,** Montevideo.
14–28 **November,** overland trip to Mercedes [discovers new fossil fragments of *Megatherium*] and returns to Montevideo.

Reconstruction of *Megatherium*

1834

[23 **December 1833–**] 4 **January 1834,** Deseado, on the Patagonian coast.
9–19 **January,** San Julián [discovers fossil bones of *Macrauchenia*].

Expedition to the Cordillera

26 January, the *Beagle* enters the Straits of Magellan.
29 January–7 March, Tierra del Fuego.
10 March–7 April, Falkland Islands.
13 April–12 May, Santa Cruz river, Patagonia.
18 April–8 May, river expeditions.
28 June–13 July, island of Chiloé, Chile.
23 July–10 November, Valparaiso.
14 August–27 September, overland expedition through the Andes.

1835

Diplolaemus darwinii.

[21 November 1834–] 4 February, Chiloé and the Chonos Islands.
8–22 February, Valdivia. [Earthquake on 20 February.]
4–7 March, Concepción.
11 March, Valparaiso.
13 March, Santiago.
18 March– 7 April, overland trip from Santiago to Mendoza, across the Andes, and return journey to Santiago, then Valparaiso.
27 April–4 July, overland trip to Coquimbo and to Copiapó, and return to the coast.
12–15 July, Iquique, Peru.
19 July–7 September, Callao, port of Lima, Peru.
16 September–20 October, Galapagos islands [important zoogeographical observations, especially of birds].
9 November, Low (or Tuamotu) archipelago [first encounter with a coral reef].
15–26 November, Tahiti.
21–30 December, New Zealand.

1836

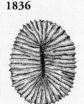

Plate coral (*Fungia*)

12–30 January, Sydney, Australia.
5–17 February, Hobart, Tasmania.
6–14 March, King George Sound, south-west Australia.
1–12 April, Cocos or Keeling Islands [coral islands; studies an atoll].
29 April–9 May, Mauritius [volcanic island].
31 May–18 June, Cape of Good Hope.
8–14 July, Saint Helena [volcanic island].
19–23 July, Ascension Island [volcanic island].
1–6 August, Bahia, Brazil.
12–17 August, Pernambuco, Brazil.
31 August, Cape Verde.
19–24 September, the Azores.
2 October, the *Beagle* docks at Falmouth, England.

Childhood tales

Darwin wrote his autobiography in 1876, explaining 'I have thought that the attempt would amuse me, and might possibly interest my children or their children.' This extract covers the days when the great naturalist was still, as he admitted, 'in many ways a naughty boy'.

By the time I went to [school] my taste for natural history, and more especially for collecting, was well developed. I tried to make out the names of plants, and collected all sorts of things, shells, seals, franks, coins, and minerals. The passion for collecting which leads a man to be a systematic naturalist, a virtuoso, or a miser, was very strong in me, and was clearly innate, as none of my sisters or brother ever had this taste.

One little event during this year has fixed itself very firmly in my mind, and I hope that it has done so from my conscience having been afterwards sorely troubled by it; it is curious as showing that apparently I was interested at this early age in the variability of plants! I told another little boy (I believe it was Leighton, who afterwards became a well-known lichenologist and botanist), that I could produce variously coloured polyanthuses and primroses by watering them with certain coloured fluids, which was of course a monstrous fable, and had never been tried by me. I may here also confess that as a little boy I was much given to inventing deliberate falsehoods, and this was always done for the sake of causing excitement. For instance, I once gathered much valuable fruit from my father's trees and hid it in the shrubbery, and then ran in breathless haste to spread the news that I had discovered a hoard of stolen fruit.

I must have been a very simple little fellow when I first went to the school. A boy of the name of Garnett took me into a cake shop one day, and bought some cakes for which he did not pay, as the shopman trusted him. When we came out I asked him why he did not pay for them, and he instantly answered, 'Why, do you not know that my uncle left a great sum of money to the town on condition that every tradesman should give whatever was wanted without payment to any one who wore his old hat and moved [it] in a particular manner?' and he then showed me how it was moved. He then went into another shop where he was trusted, and asked for some small article, moving his hat in the proper manner, and of course obtained it without payment. When we came out he said, 'Now if you like to go by yourself into that cake-shop (how well I remember its exact position) I will lend you my hat, and you can get whatever you like if you move the hat on your head properly.' I gladly accepted the generous offer, and went in and asked for some cakes, moved the old hat and was walking out of the shop, when the shopman made a rush at me, so I dropped the cakes and ran for dear life, and was astonished by being greeted with shouts of laughter by my false friend Garnett.

Charles Darwin
Autobiography, 1876

Darwin's sons

William Erasmus (1839–1914)

Darwin closely observed his firstborn son throughout his childhood. The result was the *Biographical Sketch of an Infant* which was not published until 1877. These observations were Darwin's attempt to determine and describe innate expressions and behaviour, as well as the processes of gradual acquisition. William Erasmus became a banker in Southampton.

George Howard (1845–1912)

Astronomer and mathematician, he became a Fellow of the Royal Society in 1879. In 1883, after his father had died, he was named Professor of Astronomy and Experimental Philosophy at Cambridge University, a post he occupied until his own death. He worked on the origin and evolution of the solar system, and helped his father with statistical research into marriages between blood relations.

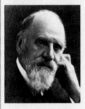

Francis (1848–1925)

A botanist specializing in plant physiology. He worked with his father from 1875 onwards, in particular on the writing of *The Power of Movement in Plants* (1880). He also wrote a biography of his father and edited part of his correspondence [*Life and Letters of Charles Darwin*, 1887, and *More Letters of Charles Darwin*, in collaboration with A. C. Seward, 1903]. Elected a Fellow of the Royal Society in 1879, he also taught at Cambridge from 1884 onwards, where he held the chair of Professor of Botany (1888–1904).

Leonard (1850–1943)

A soldier in the Royal Engineers (1871), and a major from 1890, he taught at the School of Military Engineering in Chatham (1877–82), served in the Ministry of War (Intelligence, 1885–90), and later became a liberal-unionist member of parliament for the constituency of Lichfield in Staffordshire (1892–95). He was President of the Royal Geographic Society from 1908 to 1911. A eugenist in the tradition of Galton, and a 'social Darwinist' through his strict opposition to aid for the poor (which, conversely, his father had believed in and practised), Leonard personified the distortions, misunderstandings and simplifications that have often dominated the interpretation of Darwin's thought.

Horace (1851–1928)

An engineer, and builder of scientific instruments, in 1885 he founded the Cambridge Scientific Instrument Company. Mayor of Cambridge 1896–97. Fellow of the Royal Society in 1903. Father of the future Nora Barlow, who edited several texts by her grandfather (*Diary of the Beagle*, 1933; *Charles Darwin and the Voyage of the Beagle*, 1945; *Autobiography*, 1958; *Ornithological Notes*, 1963; correspondence between the young Darwin and Henslow, 1967). She died at the age of 104.

FURTHER READING

WORKS BY DARWIN

1. Works published in Darwin's lifetime

Journal of Researches into the Geology and Natural History of the Various Countries visited by H.M.S. Beagle under the Command of Captain FitzRoy, R.N., from 1832 to 1836. London, 1839. This single volume is a reprint, with some very small changes, of a book published several months earlier as the third volume of FitzRoy's *Narrative*, which collected together the accounts of the *Beagle*'s last two voyages. Second reprint 1840. Revised second edition 1845.

The Zoology of the Voyage of H.M.S. Beagle under the Command of Capt. FitzRoy, R.N., during the years 1832 to 1836, Part I, Fossil Mammalia, by Richard Owen. London, 1840.

The Zoology of the Voyage of H.M.S. Beagle under the Command of Capt. FitzRoy, R.N., during the years 1832 to 1836, Part II, Mammalia, by Robert Waterhouse. London, 1839.

The Zoology of the Voyage of H.M.S. Beagle under the Command of Capt. FitzRoy, R.N., during the years 1832 to 1836, Part III, Birds, by John Gould. London, 1841.

The Zoology of the Voyage of H.M.S. Beagle under the Command of Capt. FitzRoy, R.N., during the years 1832 to 1836, Part IV, Fish, by the Rev. Leonard Jenyns. London, 1842.

The Zoology of the Voyage of H.M.S. Beagle under the Command of Capt. FitzRoy, R.N., during the years 1832 to 1836, Part V, Reptiles, by Thomas Bell. London, 1843.

The Structure and Distribution of Coral Reefs. Being the First Part of the Geology of the Voyage of the Beagle, under the Command of Capt. FitzRoy, R.N., during the years 1832 to 1836, London, 1842; 2nd ed. 1874, 3rd ed. 1889.

Observations on the Volcanic Islands visited during the Voyage of H.M.S. Beagle, together with some Brief Notices of the Geology of Australia and the Cape of Good Hope. Being the Second Part of the Geology of the Voyage of the Beagle, under the Command of Capt. FitzRoy, R.N., during the years 1832 to 1836. London, 1844; 2nd ed. 1876, 3rd ed. 1891.

Geological Observations on South America. Being the Third Part of the Geology of the Voyage of the Beagle, under the Command of Capt. FitzRoy, R.N., during the years 1832 to 1836. London, 1846.

[These three last books were published in a single volume in 1851 under the title *Geological Observations on Coral Reefs, Volcanic Islands, and on South America*.]

A Monograph on the Sub-Class Cirripedia, with Figures of all the Species. The Lepadidae: or, Pedunculated Cirripedes, London, 1851.

A Monograph on the Fossil Lepadidae, or Pedunculated Cirripedes of Great Britain, London, 1851.

A Monograph on the Sub-Class Cirripedia, with Figures of all the Species. The Balanidae (or Sessile Cirripedes); The Verrucidae, etc., London, 1854.

A Monograph on the Fossil Balanidae and Verrucidae of Great Britain, London, 1854.

'On the Tendency of Species to Form Varieties; and on the Perpetuation of Varieties and Species by Natural Means of Selection', the first academic presentation of the theory of selection, made on 1 July 1858 by Lyell and Hooker before the Linnean Society of London. Includes two extracts from Darwin and one from Wallace. *Journal of the Proceedings of the Linnean Society of London (Zoology)*, vol. 3 (1859), p. 45–62.

On the Origin of Species by Means of Natural Selection, or the Preservation of Favoured Races in the Struggle for Life, London, John Murray, 1859. 2nd ed. 1860, 3rd ed. 1861 (includes the preliminary 'Historical Notice' listing Darwin's transformist predecessors), 4th ed. 1866, 5th ed. 1869 (first appearance of the expression 'survival of the fittest', borrowed from Spencer), 6th ed. 1872 (definitive). The initial '*On*' disappears from the title, and there is an additional glossary by W. S. Dallas, who also compiled the index. The term 'evolution' now appears twice, although Darwin had used the word for the first time in *The Descent of Man* a year earlier. The word 'evolved' is the last word in the book in all subsequent editions.

On the Various Contrivances by which British and Foreign Orchids are Fertilised by Insects, and on the Good Effects of Intercrossing, London, 1862. 2nd edition 1877.

On the Movements and Habits of Climbing Plants, London, 1865. 2nd edition 1875.

The Variation of Animals and Plants under Domestication, London, 1868, 2 vols. 2nd edition 1875.

The Descent of Man, and Selection in relation to Sex, London, 1871. 2 vols, 2nd edition 1874 (including an additional note by Huxley).

The Expression of the Emotions in Man and Animals, London, 1872. 2nd edition by Francis Darwin, 1890.

'A Biographical Sketch of an Infant', *Mind (A Quarterly Review of Psychology and Philosophy)*, No. 7, July 1877, p. 285–94. This is based on notes taken by Darwin while observing his eldest son, William Erasmus for the first two years of his life (1839–41).

Insectivorous Plants, London, John Murray, 1875. 2nd revised edition by Francis Darwin, 1888.

The Effects of Cross and Self-Fertilisation in the Vegetable Kingdom, London, 1876. 2nd edition in 1878.

The Different Forms of Flowers on Plants of the Same Species, London, 1877. 2nd edition 1878. Corrections and new preface in 1880. Preface by Francis Darwin, 1884.

'Preliminary Notice' ('Life of Erasmus Darwin') in Ernst Krause, *Erasmus Darwin* [translated from the German by W. S. Dallas], London, 1879.

The Power of Movement in Plants [in collaboration with Francis Darwin], London, 1880.

The Formation of Vegetable Mould, through the Action of Worms, with Observations on their Habits, London, 1881.

An almost complete edition of Darwin's journal articles was published in 1977 by P. H. Barrett as *The Collected Papers of Charles Darwin*, Chicago and London, 2 vols, preface by T. Dobzhansky.

The standard English edition of Darwin's works remains *The Works of Charles Darwin*, edited by Paul H. Barrett and R. B. Freeman, 1986–89, reprinted 1987–89.

2. Manuscripts published after Darwin's death

Charles Darwin's Diary of the Voyage of H.M.S. Beagle, edited by Nora Barlow, Cambridge, 1933. This is a chronological account of the voyage, which served as the basis for the *Journal of Researches*.

Charles Darwin's Notebooks, 1836–1844. Geology, Transmutation of Species, Metaphysical Enquiries. Transcribed and edited by Paul H. Barrett, P. J. Gautrey, S. Herbert, D. Kohn and S. Smith, British Museum of Natural History, Cambridge University Press, 1987. This collection completes and improves the work previously carried out by Sir Gavin De Beer in *Darwin's Notebooks on Transmutation of Species*, published in successive parts in the *Bulletin of the British Museum of Natural History* (*Historical Series*, 1960, 2; 1961, 2; 1967, 3). The modern 1987 edition begins with the *Red Notebook* (1836–37), already edited by S. Herbert in 1980, which includes a chronological list of sites visited by the *Beagle* from May to September 1836, and a second part (1837) which primarily concerns geology but also contains thoughts on ancient and modern

geographical distribution of species, as well as some reflections relevant to the new theory. Next come two collections under the title *Geology*, *Notebook* A (1837–39) and the *Glen Roy Notebook* (early summer 1838), which describes the banks of the valley of the Glen Roy river in Scotland, and its distinctive geological structure ('parallel roads') which in 1839 Darwin wrongly thought to have a marine origin. The heart of the *Notebooks*, under the title *Transmutation of Species*, is made up of theoretical views on the scientific data collected during the voyage of the *Beagle* and analysed from a transformist point of view after Darwin's return to England. *Notebooks* B (1837–38), C (1838), D (1838), E (1838–39) and the *Torn Apart Notebook* (1839–41) may well contain Darwin's earliest reflections on the 'transmutation' of species. These are followed by the notes of *Summer 1842*, dating from the time when Darwin had just sketched the first outline of his theory; and the *Edinburgh Notebook*, the only one that includes notes taken before the voyage. It is filled with zoological notes begun in 1827 at the University of Edinburgh as well as notes dating from after his return (1837–39); the *Questions and Experiments* (1839–44) close this part of the book, thematically linked to 'transmutation'. Finally, under the title *Metaphysical Enquiries* come *Notebooks* M and N, which are rather philosophical in tone (thoughts on the origin of human faculties, behaviour, thought, reason, language, expression, belief, emotions, etc.), the *Old and Useless Notes* (1838–40), and reading notes on John MacCulloch and natural theology (*Abstract of MacCulloch*, 1838).

Eighteen of Darwin's original field notebooks have been found and are preserved at Down House. Among them is the *Santiago Book*, in which he outlines the theory of the formation of coral reefs. Darwin's notes on coral islands were transcribed and published in *Atoll Research Bulletin*, no. 88, 1962. The *St Helena Model Notebook* (1838–39) is an analysis of a plaster model of the volcanic island of St Helena. The *Reading Notebooks*, first published in 1977 ('The Darwin Reading Notebooks, 1838–1860', *Journal of the History of Biology*, vol. 10, p. 107–53), then in vol. IV of the *Correspondence* published by Cambridge (see below), list the works he had read or intended to read between 1839 and 1860. A *Zoological Diary* is preserved at Cambridge.

The Foundations of the Origin of Species. Two Essays written in 1842 and 1844 by Charles Darwin. Edited by his son Francis Darwin, Cambridge, 1909. Written in pencil in a very elliptical style that is difficult for an inexperienced reader to understand, the 1842 outline is made up of 35 sheets (53 pp in the printed version) divided into two parts, the first on variation (domestic, natural and psycho-instinctual), the second on the evidence in favour of transformism (geology, biogeography, classification, unity of type, rudimentary organs). The essay of 1844 follows this same general division, but has been heavily rewritten, extending to 230 sheets (198 in the printed version).

Charles Darwin's Natural Selection, being the Second Part of his Big Book, written from 1856 to 1858, edited by R. C. Stauffer, Cambridge, 1975.

The Autobiography of Charles Darwin 1809–1882. With original omissions restored. Edited with Appendix and Notes by his grand-daughter Nora Barlow, London, 1958. Reprinted New York, London, 1993. The first edition of Darwin's *Autobiography* (written between 28 May and 3 August 1876 for his wife Emma and children, and added to during the following years) was published at the front of *The Life and Letters,* edited by Francis Darwin in 1887 (see below). Emma was behind the initial suppression of several passages which she thought potentially embarrassing or which clashed with her own beliefs; these were mainly connected with religion or specific people who were still alive at the time.

Darwin's Journal, ed. by G. R. De Beer, *Bulletin of the British Museum (Natural History), Historical Series,* vol. 2, 1959, p. 1–21.

'Darwin's Ornithological Notes', edited by Nora Barlow, *Bulletin of the British Museum (Natural History), Historical Series,* vol. 2, 1963, p. 201–78.

Preface to August Weismann's *Studies in the Theory of Descent,* London, 1882.

Preface to Hermann Müller's *The Fertilisation of Flowers,* London, translated from the German by D'Arcy W. Thompson, 1883.

'Essay on Instinct', in G. J. Romanes, *Mental Evolution in Animals. With a Posthumous Essay on Instinct by Charles Darwin,* London, 1883.

Catalogue of the Library of Charles Darwin, now in the Botany School, Cambridge. Compiled by H. W. Rutherford, with an introduction by Francis Darwin, Cambridge, 1908.

Darwin's Library: List of Books received in the University Library of Cambridge, 'March–May 1961', mimeographed.

Charles Darwin's Marginalia, edited by M. A. Di Gregorio, with the assistance of N. W. Gill, New York and London, 1990.

3. Correspondence

Francis Darwin (ed.), *The Life and Letters of Charles Darwin, including an Autobiographical Chapter,* London, 1887, 3 vols. The autobiographical chapter had several passages removed at the request of Emma Darwin (see above).

Francis Darwin and A. C. Seward (eds). *More Letters of Charles Darwin. A Record of his Work in a Series of Hitherto Unpublished Letters,* London, 1903, 2 vols.

Henrietta Emma Litchfield, *Emma Darwin, wife of Charles Darwin. A Century of Family Letters,* Cambridge, 1904, 2 vols. Print run of 250 copies. New edition 1915.

F. Burkhardt and S. Smith (eds), *The Correspondence of Charles Darwin,* Cambridge, New York, New Rochelle, Melbourne, Sydney, Cambridge. Chronological and scholarly publication of Darwin's entire correspondence from 27 October 1821.

ON DARWIN AND DARWINIAN THEORY

Allen, Grant, *Charles Darwin,* 1886

Barlow, Nora, *Charles Darwin and the Voyage of the Beagle,* 1945

Bowlby, John, *Charles Darwin, A New Life,* 1990

Bowler, Peter J., *Charles Darwin: the Man and his Influence,* 1990

Browne, Janet, *Charles Darwin: Voyaging,* 1995

De Beer, Gavin, *Charles Darwin and Thomas Henry Huxley, Autobiographies,* 1974

De Beer, Gavin, *Charles Darwin: a Scientific Biography,* 1963

Desmond, Adrian, and Moore, James, *Darwin,* 1991

Freeman, R. B., *The Works of Charles Darwin: an annotated bibliographical handlist* (2nd ed.), 1977

Freeman, R. B., *Charles Darwin: a Companion,* 1978

Ghiselin, Michael T., *The Triumph of the Darwinian Method* (2nd ed.), 1984

Gillespie, Neal C., *Charles Darwin and the Problem of Creation,* 1979

Glass, Bentley, Temkin, Owsei and Straus, Jr., William L., *Forerunners of Darwin: 1745–1859* (2nd ed.), 1968

Glick, Thomas F., *The Comparative Reception of Darwinism,* 1972

Gruber, Howard E., and Barrett, Paul H., *Darwin on Man: a Psychological Study of Scientific Creativity, by Howard E. Gruber, together with Darwin's Early and Unpublished Notebooks*

Transcribed and Annotated by Paul H. Barrett, 1974

Hull, David L., *Darwin and his Critics: the Reception of Darwin's Theory of Evolution by the Scientific Community,* 1973

Huxley, Julian, and Kettlewell, H. B. D., *Charles Darwin and his World,* 1965

Larson, Edward J., *Summer for the Gods: The Scopes Trial and America's Continuing Debate over Science and Religion,* New York, 1997

Kohn, David (ed.), *The Darwinian Heritage,* 1985

Mayr, Ernst, *One Long Argument: Charles Darwin and the Genesis of Modern Evolutionary Thought,* 1991

Moorehead, Alan, *Darwin and the Beagle,* 1969

Peckham, Morse, *The Origin of Species by Charles Darwin: a Variorum Text,* 1959

Poulton, Edward B., *Darwin and the Emergence of Evolutionary Theories of Mind and Behavior,* 1987

Ridley, Mark, *The Essential Darwin,* 1987

Seward, A. C., *Darwin and Modern Science: Essays in Commemoration of the Centenary of the Birth of Charles Darwin and of the Fiftieth Anniversary of the Origin of Species,* 1909

Tort, Patrick, *La Pensée hiérarchique et l'évolution,* 1983

Tort, Patrick (ed.), *Darwinisme et société,* 1992

Tort, Patrick (ed.), *Dictionnaire du darwinisme et de l'évolution,* 1996

Tort, Patrick, *Darwin et le darwinisme,* 1997

Tort, Patrick (ed.), *Pour Darwin,* 1997

Tort, Patrick, *Spencer et l'évolutionnisme,* 1997

Vorzimmer, Peter J., *Charles Darwin: the Years of Controversy: The Origin of Species and its Critics 1859–1882,* 1970

Wallace, Alfred Russel, *My Life: a Record of Events and Opinions,* 1905

White, Michael and Gribbin, John, *Darwin: A Life in Science,* 1995

DOWN HOUSE

Darwin's family home and gardens are open to the public, and now house an exhibition about his life and work.

Address: Down House, Luxted Road, Downe, Kent, England, BR6 7JT

For tickets and visiting information, contact English Heritage.

Tel.: +44 (0)870 603 0145

Website: http://www.english-heritage.org.uk/

LIST OF ILLUSTRATIONS

The following abbreviations have been used:
a above; **b** below; **c** centre; **l** left; **r** right.

COVER

Front Charles Darwin, as page 92; Death's head hawkmoth, as page 4; Skeletons, detail from "The Evolution of Homo Sapiens," by Maurice Wilson in Sir Wilfred Le Gros Clarke, *History of the Primates,* 1970. By permission of the Natural History Museum, London
Spine Monkeys, from Buffon, *Histoire Naturelle, Mammifères,* 1837
Back Turtles, as page 1

OPENING

1–9 Natural history plates, coloured engravings. Private collection.
1a Spiny softshell turtle (*Apalone spinifera*)
1b Green turtle, (*Chelonia mydas*)
2 Dauw or mountain zebra
3 Shorthorn cow
4 Death's head hawkmoth (*Acherontia atropos* L.)
5 Pair of stag beetles with other beetle species

6 Lesser bird of paradise (*Paradisea minor*)
7 Belgian carrier pigeon
8 Orchid
9 European dodder
11 Photograph of Charles Darwin, 1854

CHAPTER 1

12 Charles Darwin, watercolour by George Richmond, 1840. Down House, Downe, Kent
13 Darwin's microscope. Down House, Downe, Kent
14 Erasmus Darwin, painting by Joseph Wright of Derby, *c.* 1770. Darwin College, Cambridge
14–15 The Wedgwood family, by George Stubbs, *c.* 1780. Wedgwood Museum, Barlaston
15 Josiah Wedgwood, engraving by W. Holl, *c.* 1765. The Hulton Getty Picture Collection
16 The library at Shrewsbury School, engraving, from C. W. Radclyffe, *Memorials of Shrewsbury School,* 1843
16–17 Charles Darwin and his sister Catherine, pastel drawing by Rolinda Sharples, *c.* 1816. Down House, Downe, Kent
18 Robert Waring Darwin, painting. Down House, Downe, Kent

INDEX

ACKNOWLEDGMENTS

The author would like to thank Aurélien Berra, Michel Boulard, Jean-Pierre Gasc, Daniel Goujet, Jacqueline Goy, Monique Guibert, Goulven Guilcher, Lionel Jobert, Edmée Ladier, Giovanni Landucci, Guillaume Lecointre, Gerard Morel, Musée d'Histoire Naturelle de Montauban, Florimond Parent, Claude Rives, Sacha Strelkoff, Raymond Tardy, Anne Toulemont.

PHOTO CREDITS

Patrick Tort is a philosopher, linguist and theoretician in biological sciences and humanities. He has written or edited some thirty books, and is the founder and director of the Institut Charles Darwin International. He was awarded a prize by the French Academy of Sciences for his *Dictionnaire du Darwinisme et de l'Evolution* (PUF 1996). In 2000 he won the Philip Morris Prize for history of science for his work. He is managing editor of a French edition of Darwin's complete works for Editions Syllepse.

'It is those who know little, and not those who know much, who so positively assert that this or that problem will never be solved by science.'
The Descent of Man., Introduction

Translated from the French by Paul G. Bahn

For Harry N. Abrams Inc.
Editor: Eve Sinaiko
Typographic designers: Elissa Ichiyasu, Tina Thompson
Cover designer: Brankica Kovrlija

Library of Congress Cataloging-in-Publication Data

Tort, Patrick.
    [Darwin et la science de l'évolution. English]
    Darwin and the Science of evolution / Patrick Tort.
        p. cm. - (Discoveries)
    Includes bibliographical references (p.) and index.
    ISBN 0-8109-2136-7 (pbk.)
    1. Darwin, Charles 1809-1882. 2. Naturalists-England-Biography.
    3. Evolution (Biology)   I. Title. II. Discoveries (New York, N.Y.)

    QH31.D2.T6713 2001
    576.8'092-dc21   2001022410

Copyright © 2000 Gallimard

English translation © 2001 Thames & Hudson Ltd, London

Published in 2001 by Harry N. Abrams, Incorporated, New York

Printed and bound in Italy by Editoriale Lloyd, Trieste.